Nano-Optics and Nanophotonics

Editor-in-Chief

Motoichi Ohtsu, Research Origin for Dressed Photon, Kanagawa, Japan

Series Editors

Ariando, Department of Physics, National University of Singapore, Singapore, Singapore

Sonia Contera, Department of Physics, University of Oxford, Oxford, UK

Chennupati Jagadish, Research School of Physics, Australian National University, Canberra, ACT, Australia

Fedor Jelezko, Institut für Quantenoptik, Universität Ulm, Ulm, Baden-Württemberg, Germany

Gilles Lerondel, ICD/CNRS—UMR STMR 6281, Université de Technologie de Troyes, Troyes Cedex, France

Hitoshi Tabata, Graduate School of Engineering, The University of Tokyo, Tokyo, Japan

Peidong Yang, College of Chemistry, University of California, Berkeley, CA, USA

Gyu-Chul Yi, Department of Physics, Seoul National University, Seoul, Korea (Republic of)

The Springer Series in Nano-Optics and Nanophotonics provides an expanding selection of research monographs in the area of nano-optics and nanophotonics, science- and technology-based on optical interactions of matter in the nanoscale and related topics of contemporary interest. With this broad coverage of topics, the series is of use to all research scientists, engineers, and graduate students who need up-to-date reference books. The editors encourage prospective authors to correspond with them in advance of submitting a manuscript. Submission of manuscripts should be made to the editor-in-chief, one of the editors, or to Springer.

Motoichi Ohtsu · Hirofumi Sakuma

Dressed Photons to Revolutionize Modern Physics

Exploring Longitudinal Electromagnetic Waves and Off-Shell Quantum Fields

Motoichi Ohtsu
Department of Research
Research Origin for Dressed Photon
Yokohama, Japan

Hirofumi Sakuma
Department of Research
Research Origin for Dressed Photon
Yokohama, Japan

ISSN 2192-1970 ISSN 2192-1989 (electronic)
Nano-Optics and Nanophotonics
ISBN 978-3-031-77943-5 ISBN 978-3-031-77944-2 (eBook)
https://doi.org/10.1007/978-3-031-77944-2

© The Editor(s) (if applicable) and The Author(s), under exclusive license to Springer Nature
Switzerland AG 2025

This work is subject to copyright. All rights are solely and exclusively licensed by the Publisher, whether
the whole or part of the material is concerned, specifically the rights of translation, reprinting, reuse
of illustrations, recitation, broadcasting, reproduction on microfilms or in any other physical way, and
transmission or information storage and retrieval, electronic adaptation, computer software, or by similar
or dissimilar methodology now known or hereafter developed.
The use of general descriptive names, registered names, trademarks, service marks, etc. in this publication
does not imply, even in the absence of a specific statement, that such names are exempt from the relevant
protective laws and regulations and therefore free for general use.
The publisher, the authors and the editors are safe to assume that the advice and information in this book
are believed to be true and accurate at the date of publication. Neither the publisher nor the authors or
the editors give a warranty, expressed or implied, with respect to the material contained herein or for any
errors or omissions that may have been made. The publisher remains neutral with regard to jurisdictional
claims in published maps and institutional affiliations.

This Springer imprint is published by the registered company Springer Nature Switzerland AG
The registered company address is: Gewerbestrasse 11, 6330 Cham, Switzerland

If disposing of this product, please recycle the paper.

Preface

Dressed photon (DP) is a quantum field that mediates the interaction between nanometer-sized particles (NPs). It localizes at an NP and its size is much smaller than the wavelength of a propagating light (a free photon). That is, DP is not an entity directly observed by conventional methods in optics and it should not be confused with a special type of linear evanescent light field, because, through previous research on *optical near field*, it has been found that DP is generated by nonlinear field interactions between matter and incident light field, for which quantum off-shell momentum field plays an important role. DP has unique features that are complimentary to those of the free photon (on-shell field), and a variety of novel phenomena originate from DP. Since they are analogous to some physical, chemical, and biological phenomena, we conjecture that those similarities are the manifestation of underlying mathematical universality represented by nonlinear off-shell field interactions.

By the authors' previous publications, the results of experimental studies and their application to innovative technologies have been reviewed. Although some theories were also introduced, they were prototypes built by modifying conventional on-shell scientific theories. Even though this introduction seemed to be successful in analyzing some experimental results, the problem was that these theories did not deal with the concept of "field interaction" mentioned above in a satisfactory fashion.

However, in the last few years, theoretical studies on off-shell science have rapidly progressed to solve this problem. Based on this progress, this book reviews the theories of DP creation and relevant phenomena. The first half of this book introduces the results of experimental studies, application technologies, prototype theories, and their problems: Chap. 1 presents fifteen novel phenomena originating from DP and reviews the history of DP studies. Chapter 2 demonstrates application technologies based on unique features of DP. These technologies are complimentary to those of the conventional ones. Chapter 3 reviews the experimental grounds of the unique phenomena of DP energy transfer. Numerical simulation is also reviewed that was carried out by prototype methods modifying on-shell scientific approaches, and their problems are pointed out. Chapter 4 describes numerical simulation based on a quantum walk model that was developed to solve these problems.

The second half of this book deals with the recent theoretical progress on off-shell field study focusing on DP dynamics. In the first half of this Preface, we have emphasized the important role played by nonlinear field interactions between matter and light fields in DP dynamics. As Einstein's theories of special and general relativity clearly show, light field as free electromagnetic one is closely related to what we call "physical space-time". From such a viewpoint, "matter and light field" interactions may be formally regarded as "matter and physical space-time" interactions just like matter and gravitational field interactions. As we know, the notions of space and time were originally introduced into a given physical system under consideration as purely mathematical quantities called coordinates. Although the above-mentioned Einstein's theory had revolutionized the situation, we can safely say that the present status of "physical space-time" is not complete, as is typically shown by the presence of cosmological term $\Lambda g_{\mu\nu}$ in Einstein's field equation. Occasionally as has been ridiculed by the term *Einstein's mollusk*, metric tensor $g_{\mu\nu}$ itself is not a physical quantity.

The most important aspect of our accomplishment we are going to explain in the second half is that we have succeeded in formulating a complete theory on "physical space-time" which covers not only timelike but also spacelike components of it by utilizing Greenberg-Robinson theorem in the axiomatic quantum field theory. Presumably, we can say that the problems of the unification of four forces, dark energy, and dark matter stand as big three enigmas in the contemporary theoretical physics. We believe that the reason why we cannot understand dark energy and matter is because we do not have a relevant theory on "physical space-time".

The second half starts from Chap. 5, which gives introductory remarks on the following Chaps. 6–10. Since the knowledge on Hamiltonian structure of the classical physics plays a quite important role in formulating our notion of "physical space-time", Chap. 6 is reserved for the explanation of it. In Chap. 7, we discuss several cutting-edge topics including dark energy, which are related to the spacelike part of "physical space-time". By combining the notions of conformal gravity and of the timelike part of "physical space-time", we are going to solve the mystery of dark matter in Chap. 8. The noticeable advantage of our new form of conformal gravity is the fact that it naturally bears the characteristics of spin-network as well as an entropy field different from the one in thermodynamics.

Based on the important outcomes explained in Chaps. 7 and 8, we will discuss novel cosmology in Chap. 9, and in the final Chap. 10, we will touch on a certain aspect of hierarchy problem in elementary particle physics relating to the unification of four forces, together with a couple of intriguing implications noticed on the relation between our novel cosmology and remarkable predictions of superstring theory made by Witten and Maldacena.

The first half was mainly written by M. O., the first author. The second half was mainly by H. S. However, they completed the manuscript in close cooperation with one another. The authors hope that this article will stimulate readers to gain an interest in off-shell science and to expand the routes available for reaching new studies of modern science.

Yokohama, Japan

Motoichi Ohtsu
Hirofumi Sakuma

Contents

1 Historical Review of Dressed Photon 1
  1.1 Dressed Photon, What? 1
  1.2 History 3
  References 6

2 Progresses in Experimental Studies on Dressed Photon 9
  2.1 On-Shell and Off-Shell 9
  2.2 Creation and Detection of Dressed Photons 11
  2.3 Nanofabrication Technology 13
    2.3.1 Technology Using a Fiber Probe or an Aperture 13
    2.3.2 Technology that Uses Neither Fiber nor Aperture 15
  2.4 Silicon Light-Emitting Devices 17
    2.4.1 Silicon Light-Emitting Diodes 17
    2.4.2 Silicon Lasers 20
    2.4.3 SiC Polarization Rotators 21
  References 23

3 Preliminary Theoretical Studies and Numerical Simulations 27
  3.1 Conventional Theoretical Studies on the Dressed Photon
     and Their Problems 27
    3.1.1 Creating the Dressed Photon and Coupling
       with Phonons 27
    3.1.2 Localization of the Dressed Photon 28
    3.1.3 Theoretical Problems and the Road to a Solution 29
  3.2 Spatial Evolution of DP Energy Transfer 29
    3.2.1 Size-Dependent Resonance and Hierarchy 29
    3.2.2 Autonomy 31
    3.2.3 Energy Disturbance 33
  3.3 Numerical Simulations and Their Problems 37
  References 38

4 A Quantum Walk Model for the Dressed Photon Energy Transfer 41

4.1	A Quantum Walk Model	41
4.2	Equations for the Two-Dimensional Quantum Walk Model	42
4.3	Dressed-Photon-Phonon Creation Probability on the Tip of a Fiber Probe	45
	4.3.1 Dependence on Parameters	46
	4.3.2 Dependence on the Apex Angle of a Fiber Probe	48
4.4	Dressed-Photon-Phonon Confined by a B Atom-Pair in a Si Crystal	49
	4.4.1 Dependence on the Direction of the B Atom-Pair	51
	4.4.2 Asymmetric Distribution and Photon Breeding	52
	4.4.3 A Quantum Walk Model with Energy Dissipation	53
4.5	Photon Breeding with Respect to Photon Spin	57
	4.5.1 Three-Dimensional Quantum Walk Model	58
	4.5.2 Degree of Photon Breeding	60
References		64

5 Introductory Remarks on Theoretical Chapters 6–10 65

5.1	On Off-Shell Quantum Fields	65
5.2	On the Prerequisite Knowledge	68
References		68

6 Brief Review on Generalized Hamiltonian Structure 69

6.1	Hamiltonian Structure of Particle and Perfect Fluid Systems	69
6.2	On Clebsch Parameterization of Barotropic Fluid	72
References		74

7 Off-Shell Electromagnetic Field 77

7.1	Brief Review on Free Electromagnetic Field	77
7.2	Clebsch Dual Electromagnetic Field	81
7.3	Majorana Field and DP	84
7.4	de Sitter Space and Dark Energy	86
7.5	On the Quantization of de Sitter Space	93
References		94

8 Novel Aspect of Conformal Gravity 95

8.1	Non-relativistic Representation of Ertel's Potential Vorticity	95
8.2	Converted Form of the Relativistic Equation of Motion and of PV	98
8.3	On Gravitational Entropy and Dark Matter	102
References		103

Contents

9 Novel Cosmology to be Opened up by Off-Shell Science 105
  9.1 Brief Summary on Our New Studies Explained So Far 105
  9.2 On the Meaning of DP Constant 107
  9.3 Twin Structure of the Universe 108
  9.4 On Spontaneous Conformal Symmetry Breaking of Light
      Fields 110
  References 111

10 Implications of the Novel Cosmology 113
  10.1 On Hierarchy Problem in Particle Physics 113
  10.2 On Maldacena (AdS/CFT) Duality 116
  References 118

Chapter 1
Historical Review of Dressed Photon

Abstract This chapter starts from reviewing a primitive question "What is the dressed photon (DP)?" Fifteen novel phenomena originating from the DP are presented to support this review. They are complimentary to common views that have been accepted in conventional optical science. Next, the long history of the DP studies is reviewed that originated from the studies on optical near fields. This review includes the reasons why the name "optical near field" has been changed to "dressed photon". Portions of Chap. 1 have been reproduced from Ref. [3] with permission from Elsevier, and from Ref. [4] (an open access article published under a CC BY license (Creative Commons Attribution 4.0 International License)).

1.1 Dressed Photon, What?

Dressed photon (DP) can be interpreted as a small particle of light. Unique features of the DP have been found by the experimental studies that have been carried out by the first author (M.O.) of this book [1–3]. Their results have been applied to develop varieties of revolutionary technology. However, the novel phenomena (Table 1.1) originated from the DP and the principles of these technologies might not be popularly known so far. This might be because these phenomena are complimentary to the common views that have been accepted for a long time in conventional optical science (Table. 1.2). In other words, the phenomena in Table. 1.1 and the common view in Table 1.2 are classified into *off-shell* and *on-shell* science, respectively. The off-shell science has not yet been popularly known in spite that on-shell science has been intensively studied for a long time to establish the foundation of modern science. Thus, one may arise the question "What is the DP?" at seeing the unacceptable phenomena in Table 1.1, and, finally, may lose interest in further studies of DP.

In order to answer the question above, the primary objective of this book is to review the creation mechanism and the natures of DP. After reviewing the long history of the DP studies in Sect. 1.2, Chap. 2 demonstrates the novel phenomena of Table 1.1. Chapter 3 reviews preliminary theoretical studies and their problems. Chapter 4 presents a quantum walk model and numerical simulations on the DP energy transfer. As the main topics of this book, Chaps. 5–10 theoretically review

© The Author(s), under exclusive license to Springer Nature Switzerland AG 2025
M. Ohtsu and H. Sakuma, *Dressed Photons to Revolutionize Modern Physics*,
Nano-Optics and Nanophotonics, https://doi.org/10.1007/978-3-031-77944-2_1

Table 1.1 Novel phenomena originating from DPs

No.	Phenomenon
1	The off-shell field is created and localized on a sub-wavelength material
2	The DP energy transfers back and forth between the two nanometer-sized particles (nano-particles: NPs)
3	The DP field is conspicuously disturbed and demolished by the insertion of NP2 for measurement
4	The efficiency of the DP energy transfer between the two NPs is the highest when the sizes of the NPs are equal
5	An electric-dipole-forbidden transition is allowed
6	The DP energy autonomously transfers among NPs
7	The DP energy transfer exhibits hierarchical features
8	The irradiation photon energy $h\nu$ can be lower than the excitation energy of the electron E_{excite}
9	The maximum size $a_{(DP,Max)}$ of the DP is 40–70 nm
10	The DP is created and localized at a singularity such as a nanometer-sized particle or impurity atom in a material
11	The spatial distribution of B atoms in the Si crystal varies and autonomously reaches a stationary state due to DPP-assisted annealing, resulting in strong light emission
12	The length and orientation of the B atom-pair in the Si crystal are autonomously controlled by DPP-assisted annealing
13	A light-emitting device fabricated by DPP-assisted annealing exhibits photon breeding (PB) with respect to the photon energy; i.e., the emitted photon energy $h\nu_{em}$ is equal to the photon energy $h\nu_{anneal}$ used for the annealing
14	By DPP-assisted annealing, a Si crystal works as a high-power light emitting device even though it is an indirect transition-type semiconductor
15	The semiconductor SiC crystal behaves as a ferromagnet as a result of the DPP-assisted annealing and exhibited a gigantic magneto-optical effect in the visible region

the complementarity, mechanism, and natures above, based on the concepts of off-shell science. It should be pointed out that the theories in Chaps. 5–10 also enable discussions that connect the DP in a nanometer-sized space with the cosmology in a large-scaled space.

Table 1.2 Common views in conventional optical science

No.	Phenomenon
a	Light is a propagating wave that fills a space. Its spatial extent (size) is much larger than its wavelength
b	Light cannot be used for imaging and fabrication of sub-wavelength sized materials. Furthermore, it cannot be used for assembling and operating sub-wavelength sized optical devices
c	For optical excitation of an electron, the irradiation photon energy must be equal to or higher than the energy difference between the relevant two electronic energy levels
d	An electron cannot be optically excited if the transition between the two electronic energy levels is electric-dipole forbidden
e	Crystalline silicon has a very low light emission efficiency, and thus, it is unsuitable for use as an active medium for light emitting devices

1.2 History

The new optical science of dressed photons (DPs) has seen rapid progress recently. The DP is a quantum field created in a complex system composed of photons and electrons (or excitons) in a nanometer-sized particle (nano-particle: NP). The fruits of this science have been applied to develop generic technologies (for example, nanometer-sized optical devices, information processing systems using these devices, nano-fabrication technology, and energy conversion technology) to realize disruptive innovations. Furthermore, studies on off-shell science have commenced. Off-shell science is a novel optical science including studies on the DP. The origin of this science can be found in near-field optics [3, 4].

After a long period of incubation since the 1920s, studies on near-field optics started with the aim of achieving disruptive innovations in optical science, especially in optical microscopy. Although extensive studies were carried out in the 1980s and 1990s, they have mostly already come to an end. However, basic studies exploring the nature of the optical near field (ONF) were steadily continued by a small number of scientists, and near-field optics was reincarnated as a novel science of the DP. This science involves the study of light-matter interactions in a nanometer-sized space and explores novel applications that are contrary to those in conventional optical science and technology. The first author of this book (M.O.) has been engaged in work on near-field optics for over three decades, spurred by a simple and intuitive desire to miniaturize the dimensions of light [5, 6]. He pioneered DP science based on the long-term accumulation of his studies.

First, the principles of creating and measuring the ONF should be explained: Scattered light is created when a nano-particle (NP1) is illuminated by light (Fig. 1.1a). It should be noted that another form of electromagnetic field is also created in NP1 or on its surface. This field is called the ONF. The ONF is localized on NP1, and its spatial extent (size) is equivalent to the size of NP1. The ONF cannot be measured by a conventional photodetector installed far from NP1 because it does not propagate to the far field. To measure it, a second nano-particle (NP2) is inserted into the ONF (Fig. 1.1b). The ONF is disturbed by NP2 and is converted to scattered light that propagates to the far field, and is thus measured by a photodetector.

Since the size of the ONF is equivalent to the size of NP1, it is expected that one can use the ONF to break through the diffraction limit which determines the spatial resolution in optical microscopy. In such a system, the ONF on NP1 works as a light source for acquiring an optical microscope image of NP2. Based on this expectation, research on near-field optics was started with the aim of realizing this breakthrough, and a great deal of effort was made to create an ONF whose size Δx is much smaller than the wavelength λ of light ($\Delta x \ll \lambda$).

Since conventional optical theories were used in the early studies on near-field optics, the momentum p of the electromagnetic field has been treated as a definite quantity even though it is accompanied by a small uncertainty Δp due to quantum fluctuations. However, it should be pointed out that Heisenberg's uncertainty principle $\Delta p \cdot \Delta x \geq h/2\pi$, (where h is Planck's constant) indicates a large uncertainty Δp ($\Delta p \gg p$) because of the relation $\Delta x \ll \lambda$ above.

Modern studies are treating the ONF as a quantum field with a large energy uncertainty ΔE as well as a large Δp. In particular, by examining the light-matter interactions in nanometer-sized spaces, a variety of novel phenomena that are contrary to those in conventional optical phenomena have been discovered. That is to say, near-field optics was reincarnated as a novel optical science, and the ONF was renamed the DP.

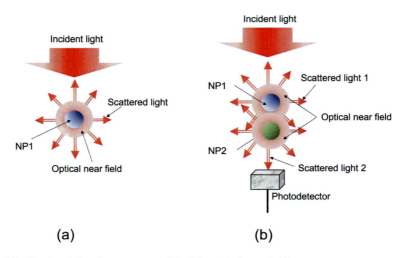

Fig. 1.1 Creation (**a**) and measurement (**b**) of the optical near field

1.2 History

It has been recognized that the classical period of near-field optics started in the 1920s to 1950s, with proposals that it be applied to high-resolution optical microscopy to achieve resolutions beyond the diffraction limit [7–9]. Diffraction and radiation of the electromagnetic field from a small aperture were analyzed based on classical electromagnetic field theory [10, 11]. Instead of using NP1 mentioned above, a small aperture was used for experimental demonstrations in the microwave frequency region.

In the 1980s and 1990s, experimental demonstrations in the optical frequency region were finally made by several scientists around the world, including the author of this book (M.O.). These include a method of acquiring an optical microscope image of sub-wavelength sized materials by scanning an aperture or a fiber probe (refer to Fig. 2.2b in Chap. 2). The equipment assembled for this imaging was named an ONF microscope, a scanning near-field optical microscope, or a near-field scanning optical microscope [12]. It was used for acquiring high-resolution images beyond the diffraction limit [13–16]. In the early stages of these studies, since the performance of the aperture or probe was not sufficiently high, it was not straightforward to acquire sufficiently high-resolution images. However, the advent of high-quality fiber probes enabled high-resolution imaging and quantitative analysis of the acquired images [17]. High-precision technology for fabricating and using high-quality fiber probes propelled the development of ONF microscopy, allowing it to join the family of scanning tunneling microscopy (STM) [18–23]. In parallel with these experimental studies, the ONF microscopes mentioned above were used to acquire images of chemical and biological specimens [24–26] and to analyze the optical properties of materials [27, 28]. These microscopes are now commercially available [29]. Evaluation of the spatial resolution of such microscopes [30], spectral analysis of the ONF [31], and a proposal for a self-consistent theory including many body effects [32] have also been made.

For promoting studies on near-field optics, a compact international workshop was held in 1992 [12]. Theoretical pictures of the ONF were drawn in this workshop based on the conventional optical method using the dispersion relation between the momentum and energy of light. Based on future outlooks for near-field optics in a panel discussion at this workshop chaired by the author (M.O.), an international near-field optics conference was organized, which is periodically held even now. To promote studies also in the Asia-Pacific region, the author (M.O.) organized the Asia-Pacific near-field optics workshop in 1996 [33]. After this workshop, the Asia-Pacific near-field optics conference was organized and is still periodically held.

In the studies above, since the ONF and relevant phenomena have been discussed in the framework of conventional wave-optics, the electromagnetic origin of the ONF, and especially the origin of the light-matter interactions in nanometer-sized space, have remained unrevealed. However, a salvation was that quantum theoretical studies were started as a first step in revealing the nature of these interactions [34, 35].

Although near-field optics had succeeded in breaking through the diffraction limit by the 1990s, a technical problem was that the data acquisition time was too long because the fiber probe had to be slowly scanned under precise feedback servo-control. Furthermore, since other members of the STM family (such as atomic force microscopes and scanning electron tunneling microscopes) had already realized spatial resolutions as high as or higher than that of ONF microscopes, it was not straightforward to make the prominent performance of ONF microscopes appealing to nonprofessional users. A more essential problem was that the image-acquiring process in an ONF microscope disturbed the electron energies in the specimen. This is because the image is acquired by using the near-field optical interaction between the tip of the fiber probe and the specimen. In other words, NP1 (the tip of the fiber probe) and NP2 (the specimen) are not independent of each other but are combined via the near-field optical interaction. Thus, even though a high spatial resolution beyond the diffraction limit was realized, the problem was that the profile of the acquired image did not have a direct correlation with that of a conventional optical microscopic image. On the whole, the classical studies of the ONF in the 1980s and 1990s did not give any clear answers to the essential questions, "What is the origin of the near-field optical interaction?" and "What kind of optical scientific revolution could near-field optics have made?" By recognizing that these questions had been neglected, the application of the ONF to microscopy, i.e., the study of near-field optics in the classical period, effectively came to an end.

However, even after this end, basic studies on the concepts and principles of the ONF were steadily continued by a small number of scientists. In relation to these studies, experiments on controlling atomic motion with high spatial resolution were carried out in a high vacuum [36]. Thanks to these continuous studies, a modern period of studies has started via transient studies named nano-optics or nanophotonics [37, 38]. As a result, novel optical science and technology, based on the DP, have emerged as the reincarnation of classical near-field optics.

References

1. Ohtsu, M.: Dressed Photons, pp. 89–214. Springer, Heidelberg (2014)
2. Ohtsu, M.: Silicon Light-Emitting Diodes and Lasers. Springer, Heidelberg (2016)
3. Ohtsu, M.: Off-Shell Applications in Nanophotonics. Elsevier, Oxford (2021)
4. Ohtsu, M.: History, current developments, and future directions of near-field optical science. Opto-Electron. Adv. **3**, 190046 (2020)
5. Ohtsu, M.: Introduction. In: Ohtsu, M. (ed.) Near-Field Nano/Atom Optics and Technology, pp. 1–14. Springer, Tokyo (1998)
6. Ohtsu, M.: Overview. In: Zhu, X., Ohtsu, M. (eds.) Near-Field Optics: Principles and Applications, pp. 1–8. World Scientific, Singapore (2000)
7. Ohtsu, M.: From classical to modern near-field optics and the future. Opt. Rev. **21**, 905–910 (2014)
8. Synge, E.H.: A suggested method for extending microscopic resolution into the ultra-microscopic region. Phil. Mag. **6**, 356–362 (1928)
9. Aloysuis, J.: Resolving power of visible light. J. Opt. Soc. Am. **46**, 359 (1956)
10. Bethe, H.: Theory of diffraction by small holes. Phys. Rev. **66**, 163–182 (1944)
11. Bouwkamp, C.J.: On the diffraction of electro-magnetic waves by small circular discs and holes. Philips Res. Rep. **5**, 401–422 (1950)
12. Pohl, D.W., Courjon, D. (eds.): Near Field Optics, Kluwer (1993)
13. Pohl, D.W., Denk, W., Lanz, M.: Optical stethoscopy: image recording with resolution $\lambda/20$. Appl. Phys. Lett. **44**, 651–653 (1984)

References

14. Lewis, A., Isaacson, M., Harootunian, A., Muray, A.: Development of a 500 A spatial resolution light microscope. I. Light is efficiently transmitted through $\lambda/16$ diameter apertures. Ultramicroscopy **12**, 227–231 (1984)
15. Fischer, U.: Optical characteristics of 0.1 μm circular apertures in a metal film as light sources for scanning ultramicroscopy. J. Vac. Sci. Technol. B **3**, 386–390 (1985)
16. Betzig, E., Isaacson, M., Lewis, A.: Collection mode near-field scanning optical microscopy. Appl. Phys. Lett. **51**, 2088–2090 (1987)
17. Mononobe, S., Saiki, T.: Near-Field Nano/Atom Optics and Technology, Chs 3 and 4. (ed Ohtsu, M.). Springer Tokyo (1998)
18. Betzig, E., Trautman, J.K., Harris, T.D., Weiner, J.S., Kostelack, R.L.: Breaking the diffraction barrier: optical microscopy on a nanometric scale. Science **251**, 1468–1470 (1991)
19. Pangaribuan, T., Yamada, K., Jian, S., Ohasawa, H., Ohtsu, M.: Reproducible fabrication technique of nanometric tip diameter fiber probe for photon scanning tunneling microscope. Japan. J. Appl. Phys. **31**, L1302–L1304 (1992)
20. Malmqvist, L., Hertz, H.M.: Trapped particle optical microscopy. Opt. Commun. **94**, 19–24 (1992)
21. Van Hulst, N.F., et al.: Near-field optical microscope using a silicon-nitride probe. Appl. Phys. Lett. **62**, 461–463 (1993)
22. Zenhausern, F., Martin, Y., Wickramasinghe, H.K.: Scanning interferometric apertureless microscopy: optical imaging at 10 angstrom resolution. Science **269**, 1083–1085 (1995)
23. Guerra, J.M.: Photon tunneling microscopy. Appl. Opt. **29**, 3741–3752 (1990)
24. Betzig, E., Chichester, R.J.: Single molecules observed by near-field scanning optical microscopy. Science **262**, 1422–1425 (1986)
25. Xie, X.S., Dunn, R.C.: Probing single molecule dynamics. Science **265**, 361–364 (1994)
26. Maheswari, R.U., Mononobe, S., Tatsumi, H., Katayama, Y., Ohtsu, M.: Observation of subcellular structures of neurons by an illumination mode near-field optical microscope under an optical feedback control. Opt. Rev. **3**, 463–467 (1996)
27. Levy, J., Nikitin, V., Kikkawa, J.M., Awschalom, D.D., Samarth, N.: Femtosecond near-field spin microscopy in digital magnetic heterostructures. J. Appl. Phys. **79**, 6095–6100 (1996)
28. Saiki, T., Nishi, K., Ohtsu, M.: Low temperature near-field photoluminescence spectroscopy of InGaAs single quantum dots. Jpn. J. Appl. Phys. **37**, 1638–1642 (1998)
29. Narita, Y., Murotani, H.: Submicrometer optical characterization of the grain boundary of optically active Cr3+ doped polycrystalline Al2O3 by near-field spectroscopy. Am. Miner. **87**, 1144–1147 (2002)
30. Isaascson, M., Cline, J., Barshatzky, H.: Resolution in near-field optical microscopy. Ultramicroscopy **47**, 15–22 (1992)
31. Wolf, E., Nieto-Vesparinas, M.: Analyticity of the angular spectrum amplitude of scattered fields and some of its consequences. J. Opt. Soc. Am. **2**, 886–890 (1985)
32. Girard, C., Courjon, D.: Model for scanning tunneling optical microscopy: a microscopic self-consistent approach. Phys. Rev. B **42**, 9340–9439 (1990)
33. Eah, S.-K., Jhe, W., Saiki, T., Ohtsu, M.: Study of quantum optical effects with scanning near-field optical microscopy. In: The First Asia-Pacific Workshop on Near Field Optics (1996)
34. Kobayashi, K., Ohtsu, M.: Quantum theoretical approach to a near-field optical system. J. Microsc. **194**, 249–254 (1999)
35. Kobayashi, K., Sangu, S., Ito, H., Ohtsu, M.: Near-field optical potential for a neutral atom. Phys. Rev. A **63**, 1–9 (2001)
36. Ito, H., Nakata, T., Sakaki, K., Ohtsu, M.: Laser spectroscopy of atoms guided by evanescent waves in micron-sized hollow optical fibers. Phys. Rev. Lett. **76**, 4500–4503 (1996)
37. Ohtsu, M., Kobayashi, K., Kawazoe, T., Sangu, S., Yatsui, T.: Nanophotonics: design, fabrication, and operation of nanometric devices using optical near fields. IEEE J. Sel. Top. Quantum Electron. **8**, 839–862 (2002)
38. Ohtsu, M. (ed.): Handbook of Nano-Optics and Nanophotonics. Springer, Heidelberg (2013)

Chapter 2
Progresses in Experimental Studies on Dressed Photon

Abstract After reviewing the nature of off-shell field and creation/measurement processes of DP, nanometer-sized optical logic gate devices are demonstrated that triggered rapid development of off-shell science. Next, two technologies are reviewed that exhibit novel phenomena of Table 1.1. One is the nanofabrication technology. Its essential by-product is the successful estimation of the maximum size of the DP. The other is silicon light-emitting devices. It is pointed out that a cluster of photons emitted from these devices behaves as if it is a single photon. Portions of Chap. 2 have been reproduced from Ref. [15] (an open access article published under a CC BY license (Creative Commons Attribution 4.0 International License)), from Ref. [47] with permission from Elsevier, and from Sakuma, H., Ojima, I. and Ohtsu, M., Perspective on an Emerging Frontier of Nanoscience Opened up by Dressed Photon Studies. Nanoarchitectonics, 1, 1–23 (2024) (an open access article distributed under a CC BY license (Creative Commons Attribution 4.0 International License)).

2.1 On-Shell and Off-Shell

From the current studies of the DP as the reincarnation of classical near-field optics, novel phenomena (Table 1.1 in Chap. 1) that are complimentary to those accepted in conventional optical science have been found. In order to review the developments in studies on the DP, Table 1.2 in Chap. 1 listed five common views [a]–[e] that have been accepted for a long time in conventional optical science. The origin of these common views is attributed to the dispersion relation of the photon, which definitely fixes the relation between energy E and momentum p. In the case where light propagates in a vacuum, the dispersion relation is linear ($E = cp$, where c is the speed of light). By noting that momentum is a three-dimensional vector, this relation is geometrically represented by a circular cone. In the case of propagation in a material, this relation is geometrically represented by a paraboloid. This circular cone and paraboloid have been called the mass-shell (Fig. 2.1), and thus, propagating light is considered to be an electromagnetic field in the on-shell state ("on-shell field" for short) because it is on the mass-shell. Even though the quantum fluctuations of the light have to be taken into account, conventional optical science has treated light in the on-shell state. Thus,

© The Author(s), under exclusive license to Springer Nature Switzerland AG 2025
M. Ohtsu and H. Sakuma, *Dressed Photons to Revolutionize Modern Physics*,
Nano-Optics and Nanophotonics, https://doi.org/10.1007/978-3-031-77944-2_2

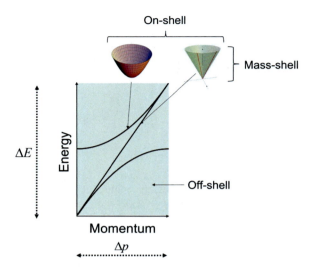

Fig. 2.1 Mass-shell (on-shell) and off-shell in the dispersion relation

this science can be called on-shell science. The common views [a]–[e] in Table 1.2 are for light in the on-shell state, and they have become accepted in on-shell science.

Complimentary to propagating light in the on-shell state described above, the ONF is in the off-shell state, which deviates from the circular cone and the paraboloid above. This is because its sub-wavelength size $\Delta x (\ll \lambda)$, being contrary to the common view [a] above, produces a large momentum uncertainty Δp ($\Delta p \gg p$) due to the Heisenberg's uncertainty principle $\Delta p \cdot \Delta x \geq h/2\pi$.

Since p and E are mutually dependent, the electromagnetic field in the off-shell state ("off-shell field" for short) also has a large uncertainty ΔE ($\gg E$) in the energy. Thus, Heisenberg's uncertainty principle ($\Delta E \cdot \Delta t \geq h/2\pi$) also gives $\Delta t \leq h/2\pi E$. This indicates the short duration of the field, which corresponds to the nature of a virtual photon. Due to the two large uncertainties (Δp and ΔE), the science of the ONF belongs to the category of off-shell science [1]. It should be noted that the nature of the off-shell quantum field is complimentary to that of the on-shell quantum field. The off-shell field is created neither in a vacuum nor in a super-wavelength sized macroscopic material. Instead, this field exhibits the following phenomenon: [Phenomenon 1] *The off-shell field is created and localized on a sub-wavelength material.*

Since this field is created as a result of the interactions between photons and electrons (or excitons) in a nanometer-sized particle (nano-particle: NP), it is the electromagnetic field that accompanies the electronic or excitonic energy. Thus, this field has been named the DP [1]. In other words, the DP is the quantum field created in a complex system composed of photons and electrons (or excitons) in an NP. It has a sub-wavelength size and short duration. By using the DP, novel phenomena that are complimentary not only to the common view [a] but also to common views [b]–[e] have been found. Preliminary theoretical studies have been carried out to draw the physical pictures of the DP (Sect. 2.2).

2.2 Creation and Detection of Dressed Photons

A DP is created at a point-like singularity, such as on a NP (Fig. 1.1a), on the tip of an optical fiber probe, on bumps of a rough material surface, or on an impurity atom in a crystal, as is schematically explained by Fig. 2.2a–d. In the case of Figs. 1.1a and 2.2a, as an example, the DP field is created in a complex system composed of photons and electrons (or excitons) in a NP. This means that the photon "dresses "the exciton energy, and thus, this field was named a DP [1].

As an example of further dressing of the material energy, coupling between a DP and a phonon has been found: The energy of the created DP transfers to the adjacent NP via a process called DP hopping. In the case where multiple NPs are the atoms in a crystal lattice (Fig. 2.2d), the DP excites a lattice vibration during the hopping, resulting in the creation of phonons. Subsequently, the DP exchanges its energy with these phonons to create a dressed-photon-phonon (DPP) quantum field [2].

To measure the DP that is created and localized on NP1, the DP must be converted to propagating scattered light. This can be performed by inserting NP2 into the DP field (Fig. 1.1b). Propagating scattered light is created by this insertion, and it reaches a photodetector in the far field, where it is measured. Although NP1 and NP2 may be considered a light source and a detector in this process, one should note the following two phenomena: [Phenomenon 2] *The DP energy transfers back and forth between the two NPs.* [Phenomenon 3] *The DP field is conspicuously disturbed and demolished by the insertion of NP2 for measurement.*

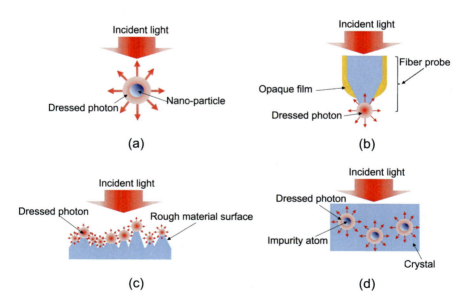

Fig. 2.2 Experimental methods for creating DPs. **a** On a NP (Fig.1.1a). **b** On the tip of a fiber probe. **c** On bumps of a rough material surface. **d** On impurity atoms in a crystal

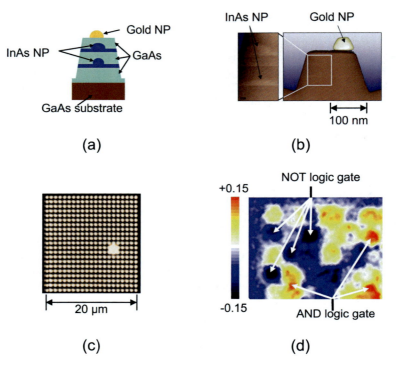

Fig. 2.3 Room temperature-operation logic gates composed of InAs NPs. **a** Cross-sectional structure of the mesa-shaped device. **b** Scanning transmission electron microscope image of the cross-sectional profile. **c** Optical microscope image of a two-dimensional array of fabricated devices. **d** Measured spatial distribution of the output signal intensity from a two-dimensional array

Furthermore, the following phenomenon was also found [3]: [Phenomenon 4] *The efficiency of the DP energy transfer between the two NPs is the highest when the sizes of the NPs are equal.* This phenomenon was named size-dependent resonance. Although the long-wavelength approximation has been popularly used in conventional optical scientific studies on light-matter interactions, it is invalid in the case of a DP because the spatial extent of a DP is much smaller than the wavelength of light. Due to this invalidity, the following phenomenon that is complementary to the common view [d] was found: [Phenomenon 5] *An electric-dipole-forbidden transition is allowed.*

The results of the above basic studies have ingeniously contributed to the realization of innovative generic technologies. For example, nanometer-sized optical functional devices were developed by using semiconductor NPs. They have enabled transmission and readout of optical signals via DP energy transfer and subsequent dissipation. Practical NOT logic gate and AND logic gate devices operated at room temperature have been fabricated by using InAs NPs (Fig. 2.3) [4].

2.3 Nanofabrication Technology

One advantage of these devices was that their extremely small sizes (20–50 nm side length in the case of the logic gate devices using InAs NPs) were far beyond the diffraction limit, complimentary to the common view [b]. Other advantages were their superior performance levels and unique functionality, such as single-photon operation [5], extremely low energy consumption [6], and autonomous energy transfer [7]. These advantages originate from the unique operating principles of DP devices achieved by exploiting Phenomena 4 and 5. Furthermore, an inherent phenomenon was used for device operation: [Phenomenon 6] *The DP energy autonomously transfers among NPs.* Novel information processing systems have been proposed by using DP devices. The first example is a non-Von Neumann computing system utilizing DP energy transfer [8]. The ability to solve a decision-making problem [9] and an intractable computational problem [10] has been demonstrated. The second example is an information security system that uses Phenomenon 4. This system has realized a lock-and-key [11]. Furthermore, a hierarchical hologram [12] has been developed using the following phenomenon that originates from the size-dependent resonance: [Phenomenon 7] *The DP energy transfer exhibits hierarchical features.*

The two sections below review two more examples of these technologies and present novel phenomena that originate from the intrinsic nature of DPs. They are complimentary to the common views [a]–[e] in Table 1.1. Table 1.1 in Chap. 1 summarized novel phenomena (including Phenomena 1–7 above) originating from DPs. Even though novel theories on light-matter interactions are required to analyze these phenomena, on-shell science has never met this requirement. However, disruptive innovations in application technologies have been realized by applying these phenomena.

2.3 Nanofabrication Technology

This section starts by reviewing an example of nanofabrication technology that uses a fiber probe (Fig. 2.2b) or an aperture for creating a DP. The specific natures of the DP relevant to this technology, that are complimentary to the common view [b], are also demonstrated. Next, a more practical technology is reviewed, in which neither a fiber probe nor an aperture is required.

2.3.1 Technology Using a Fiber Probe or an Aperture

This subsection reviews photochemical vapor deposition (PCVD) that involves molecular dissociation by a DP and subsequent deposition of the dissociated atoms on a substrate. $Zn(C_2H_5)_2$ was adopted as a specimen molecule. A DP was created on the tip of a fiber probe by irradiating the end of the fiber probe with light. Gaseous $Zn(C_2H_5)_2$ molecules, filled in a vacuum chamber, dissociated when these molecules moved into the DP field. The dissociated Zn atoms subsequently landed on a sub-

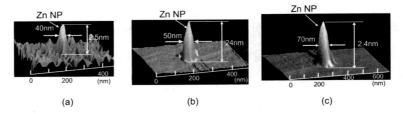

Fig. 2.4 Shear-force microscopic images of Zn NPs formed on sapphire substrates. The dissociated molecules are Zn(C$_2$H$_5$)$_2$ (**a** and **b**) Zn(acac)$_2$ (**c**). The wavelength of the propagating light for creating the DP are 684 nm (**a**), 488 nm (**b**), and 457 nm (**c**)

strate and were adsorbed on the substrate. By repeating these processes, the number of adsorbed Zn atoms increased, resulting in the deposition of Zn atoms and the formation of a nanometer-sized metallic Zn NP on the substrate.

For comparison, the wavelength in the case of dissociating Zn(C$_2$H$_5$)$_2$ molecules by using conventional propagating light had to be shorter than 270 nm (photon energy = 4.59 eV) to excite an electron in the Zn(C$_2$H$_5$)$_2$ molecule (refer to the common view [c]). By noting this requirement, the following ingenious contrivances [1] and [2] were employed to confirm that the Zn(C$_2$H$_5$)$_2$ molecules were dissociated by the above DP.

[1] The wavelength of the propagating light for creating the DP was set longer than 270 nm, complimentary to the common view [c]. However, the Zn(C$_2$H$_5$)$_2$ molecules were expected to be dissociated by the DP on the tip due to the following phenomenon: [Phenomenon 8] *The irradiation photon energy $h\nu$ can be lower than the excitation energy of the electron E_{excite}*. That is, since the created DP is the quantum field accompanying the energies of the excitons (E_{excite}) and phonons (E_{phonon}) at the tip of the fiber probe (DPP quantum field: Sect. 2.2), its energy is expressed as $h\nu_{DP} = h\nu + E_{excite} + E_{phonon}$. Thus, even though $h\nu < E_{excite}$, the DP energy $h\nu_{DP}$ can be greater than E_{excie} [13].

[2] The Zn(C$_2$H$_5$)$_2$ molecules were replaced by zinc-bis(acetylacetonate) ("Zn(acac)$_2$" for short) molecules [14]. Zn(acac)$_2$ is a well-known optically inactive molecule that has never been shown to be dissociated by propagating light. However, from Phenomenon 4, the possibility of it being dissociated by the DP was expected.

Figure 2.4a, b show images of Zn NPs formed on sapphire substrates by dissociating Zn(C$_2$H$_5$)$_2$ molecules [13]. The wavelengths of the propagating light for creating the DP were as long as 684 and 488 nm. Figure 2.4c shows an image of a Zn NP for which Zn(acac)$_2$ molecules were used [14]. The wavelength of the propagating light for creating the DP was 457 nm. Figure 2.4 demonstrates that the presented PCVD using the DP is complimentary to the common views [b]–[d].

The maximum size $a_{(DP, Max)}$ of the DP was estimated from the above experimental results [15]. For this estimation, the increasing rate R of the full-width at half-maximum (FWHM) of the formed Zn NP was measured [16]. The measured results showed that R was the maximum when the FWHM was equal to the tip diame-

2.3 Nanofabrication Technology 15

ter $2a_p$ of the fiber probe ($a_p = 4.4$ nm: tip radius). This was due to the size-dependent resonance of the DP energy transfer between the tip of the fiber probe and the formed Zn NP (Phenomenon 4). Although a further increase in the deposition time increased the FWHM, R decreased to zero. Finally, the FWHM saturated. Figure 2.4 shows the profiles acquired after this saturation.

The FWHMs in Fig. 2.4 were 40–70 nm. They were independent of the tip diameter, the wavelength and power of the light used for irradiating the end of the fiber probe, and the species of molecules used. Since the spatial profile and size of the DP transferred from the tip of the fiber probe corresponded to those of the NP deposited on the substrate, the FWHMs in Fig. 2.4 indicate the following phenomenon: [Phenomenon 9] *The maximum size $a_{(DP,Max)}$ of the DP is 40–70 nm.*

By using the above PCVD technology, a variety of two-dimensional patterns were formed by scanning a fiber probe [17]. To increase the working efficiency for pattern formation, a novel lithography technology was developed in which the fiber probe was replaced by a two-dimensional photomask [18]. A fully automatic practical photolithography machine was developed and used to form a diffraction grating pattern with a half pitch as narrow as 22 nm [19]. It also produced a two-dimensional array of the nanometer-sized optical devices reviewed in Sect. 2.1 [20] and practical devices for soft X-rays (a Fresnel zone plate [21] and a diffraction grating [22]).

2.3.2 Technology that Uses Neither Fiber nor Aperture

This subsection reviews a technology for autonomous smoothing of a material surface that requires neither fiber probes nor apertures. The material to be smoothed is installed in a vacuum chamber, and the chamber is filled with gaseous molecules. By irradiating the material surface with light, DPs are created at the tips of the bumps on the rough material surface (Fig. 2.2c). That is, the bumps play the role of fiber probes for creating DPs. If the molecules move into the DP field, they are dissociated. The chemically active atoms created as a result of this dissociation selectively etch the tips of the bumps away, while the flat part of the surface remains unchanged. The etching autonomously starts upon light irradiation, and the surface roughness gradually decreases as etching progresses. The etching autonomously stops when the bumps are annihilated and the DPs are no longer created.

The disc surface of a synthetic silica substrate (30 mm diameter) was etched by using gaseous Cl_2 molecules. Although light with a wavelength shorter than 400 nm was required for conventional photodissociation (common view [c]), the present method used visible light with a wavelength of 532 nm based on Phenomenon 8. Etching by active Cl atoms decreased the surface roughness to as low as 0.13 nm. A laser mirror was produced by coating a high-reflection film on the smoothed substrate surface. Its damage threshold for high-power ultraviolet laser light pulses was evaluated to be as high as twice that of the commercially available strongest mirror whose substrate surface was polished by a conventional chemical-mechanical polishing technology [23].

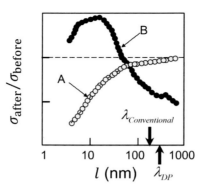

Fig. 2.5 Ratio of the standard deviation of the roughness of the PMMA surface before and after etching. Curves A and B are the results acquired by illuminating the surface with light with a wavelength of λ_{DP} = 325 nm and $\lambda_{Conventional}$ = 213 nm, respectively. The upward and downward arrows represent the values of l that are equal to the above wavelengths

Gaseous O_2 molecules can also be used for autonomous etching because the O atoms created by dissociation are chemically active. The advantage is that etching can be carried out in atmospheric conditions by using O_2 molecules in air, and thus, a vacuum chamber is not required. Curve A in Fig. 2.5 shows the experimental results of etching a plastic polymethyl methacrylate (PMMA) surface [24]. Although ultraviolet light with a wavelength shorter than 242 nm was required for the conventional photodissociation of O_2 molecules, light with a longer wavelength λ_{DP} = 325 nm was used here due to Phenomenon 8. For comparison, Curve B shows the result of etching using conventional photodissociation, for which the wavelength $\lambda_{Conventional}$ of the light used was as short as 213 nm.

In Fig. 2.5, the surface roughness was evaluated from its standard deviation $\sigma(l)$. The horizontal axis l represents the period of the roughness of the surface. The vertical axis represents the ratio $\sigma_{after}/\sigma_{before}$ between the $\sigma(l)$ values before (σ_{before}) and after (σ_{after}) etching [24]. The curve A shows that $\sigma_{after}/\sigma_{before} < 1$ in the range $l < \lambda_{DP}$, through which the contribution of the subwavelength-sized DP is confirmed. A drastic decrease in $\sigma_{after}/\sigma_{before}$ can be observed in the range $l < 40$–70 nm, which again confirms Phenomenon 9 regarding the maximum size of the DP. In contrast to the curve A, the curve B shows that $\sigma_{after}/\sigma_{before} < 1$ in the range $l > \lambda_{Conventional}$. This means that the etching was effective only in the super-wavelength range.

Since DPs are always created at the tips of the bump on the material surface under light irradiation, the present autonomous etching has been applied to smoothing of a variety of surfaces and materials: the side surface of a diffraction grating [25], the surface of a photomask used for conventional ultraviolet lithography [26], and the surfaces of GaN crystals [27], transparent ceramics [28], and diamonds [29].

2.4 Silicon Light-Emitting Devices

This section reviews silicon light-emitting diodes (Si-LEDs), Si lasers, and silicon carbide (SiC) polarization rotators. Phenomena complimentary to the common view [e] are also demonstrated.

2.4.1 Silicon Light-Emitting Diodes

Crystalline Si has long been a key material supporting the development of modern technology. However, because Si is an indirect-transition-type semiconductor, it has been considered to be unsuitable for light-emitting devices. The momentum conservation law requires an interaction between an electron-hole pair and phonons for radiative recombination. However, the probability of this interaction is very low. Nevertheless, Si has been the subject of extensive study for the fabrication of light-emitting devices [30, 31]. The above problems have been solved by using DPPs because the phonons in a DPP can provide momenta to the electron to satisfy the momentum conservation law [32, 33]. For device fabrication, DPPs were created by irradiating a Si crystal with light. For device operation, DPPs were created by electronic excitation.

For fabrication, an As atom- or Sb atom-doped n-type Si substrate was used. As the first step, the substrate surface was transformed to a p-type material by implanting B atoms, forming a p-n homojunction. Metallic films were coated on the substrate surface to serve as electrodes. As the next step, this substrate was processed by a fabrication method named DPP-assisted annealing: Joule heat was generated by current injection, which caused the B atoms to diffuse. During this Joule annealing, the substrate surface was irradiated with light (wavelength λ_{anneal} = 1.342 μm). Because its photon energy $h\nu_{anneal}$ (=0.925 eV) was sufficiently lower than the bandgap energy E_g (=1.12 eV) of Si, the light could penetrate into the Si substrate without suffering absorption. Then, the light reached the p-n homojunction to create a DPP on an impurity B atom (Fig. 2.2d). The created DPP localized at this impurity B atom, which manifested the following phenomenon: [Phenomenon 10] *The DPP is created and localized at a singularity such as a nanometer-sized particle or impurity atom in a material.*

The phonons in the created DPP could provide momenta to a nearby electron to satisfy the momentum conservation law, resulting in stimulated emission of light. The emitted light propagated from the crystal to the outside, which indicated that part of the Joule energy used for diffusing B atoms was dissipated in the form of optical energy, resulting in local cooling that locally decreased the diffusion rate. As a result, through the balance between heating via the Joule energy and cooling via the stimulated emission, the spatial distribution of B atoms varied and autonomously reached a stationary state. This stationary state was expected to be the optimum for efficient creation of DPPs and for efficient light emission because the probability of spontaneous emission was proportional to that of the stimulated emission described above. After DPP-assisted annealing, the Huang-Rhys factor, a parameter representing the magnitude of the coupling between electron-hole pairs and phonons, was experimentally evaluated to be 4.08 [34]. This was 10^2–10^3 times higher than that before DPP-assisted annealing.

The above fabricated device was operated as a light-emitting diode (LED) by simple current injection: By injecting a current of 3.0 A into the device with an areal size of 0.35 mm by 0.35 mm, a continuous wave (CW) output optical power as high as 2.0 W was obtained at a substrate temperature of 77 K. A power as high as 200 mW was obtained at room temperature (283 K) [35]. These results confirmed that the following phenomenon occurs: [Phenomenon 11] *The spatial distribution of B atoms in the Si crystal varies and autonomously reaches a stationary state due to DPP-assisted annealing, resulting in strong light emission.*

Note that the photon energy emitted from conventional LEDs is governed by E_g. However, for the present Si-LED, the light emission spectra acquired at a temperature of 283 K and an injection current of 2.45 A [35] clearly showed a high spectral peak $h\nu_{em}$ at $E_g - 3E_{phonon}$, where E_{phonon} is the phonon energy (Fig. 2.6) The origin of this spectral peak was attributed to the spatial distribution of B atoms that was autonomously controlled during DPP-assisted annealing [36]. The measured three-dimensional spatial distribution of B atoms at the p-n homojunction indicated that the B atoms were apt to form pairs with a length $d = 3a$ (a is the lattice constant of the Si crystal (=0.54 nm)) and that the formed pairs were apt to orient along a plane parallel to the top surface of the Si crystal [37]. That is, the following phenomenon was found: [Phenomenon 12] *The length and orientation of the B atom-pair in the Si crystal are autonomously controlled by DPP-assisted annealing.* Note that the required phonon momentum must be h/a for radiative recombination of the electron (at the bottom of the conduction band at the X-point in reciprocal space) and the positive hole (at the top of the valence band at the Γ-point) to occur. Since the phonon momentum is $h/3a$ when $d = 3a$, the DPP created and localized at this B atom pair provides the momenta of three phonons to the electron. As a result, $h\nu_{em}$ is expressed as $E_g - 3E_{phonon}$, and its value is 0.93 eV (E_{phonon} =65 meV), which is nearly equal to the photon energy $h\nu_{anneal}$ (=0.95 eV) irradiated during DPP-assisted annealing. This indicates that the irradiated light served as a breeder that created a photon with energy $h\nu_{em} = h\nu_{anneal}$ and manifested the following phenomenon: [Phenomenon 13] *A light-emitting device fabricated by DPP-assisted annealing exhibits photon breeding (PB) with respect to the photon energy; i.e., the emitted photon energy $h\nu_{em}$ is equal to the photon energy $h\nu_{anneal}$ used for the annealing.* PB was also observed with respect to the photon spin. That is, the polarization direction of the emitted light was identical to that of the light irradiated during DPP-assisted annealing [37].

The fabricated Si-LED was demonstrated to work as a relaxation oscillator upon injection of a direct current, yielding an emission pulse train [38]. As an advanced version of this experiment, the 2nd-order cross-correlation coefficient (CC) was measured to evaluate the photon statistical features of the emission from small light spots on the device surface, which took the form of a pulse train (duration and repetition frequency of 50 ps and 1 GHz, respectively) [39]. Figure 2.7 shows the

2.4 Silicon Light-Emitting Devices

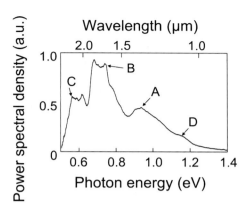

Fig. 2.6 Light emission spectra of the Si-LED. The substrate temperature and the injection current were 283 K and 2.45 A, respectively. Downward arrows A–D represent the spectral peaks at $E_g - 3E_{phonon}$, $E_g - 6E_{phonon}$, $E_g - 9E_{phonon}$, and E_g, respectively

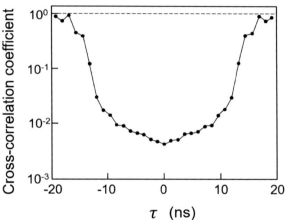

Fig. 2.7 The second-order CC measured as a function of the time difference τ of the measurements

value of the CC evaluated by a Hanbury Brown-Twiss experimental setup [40]. It presents two features. One is that the CC is smaller than unity in the range of time difference of the measurements by two independent detectors $|\tau| < 20$ ns. This indicates the photon antibunching phenomenon that is an inherent feature of a single photon. The other feature is that the CC takes a nonzero value at $\tau = 0$, although it is less than 1×10^{-2}. This small nonzero value is attributed to the photons generated from multiple light spots located in close proximity to each other on the Si-LED surface.

These two features suggest the possibility that an emitted cluster of photons behaves as if it is a single photon. This possibility can be conjectured to be related to the localizable property of the spin-zero particle, as was pointed out in [39] in relation to the Wightman theorem [41]. Namely, if the observable positions of given spin-zero quantum particles are sufficiently close, then the cluster of these particles would behave as if it is a single quantum particle with the accumulated amount of energy.

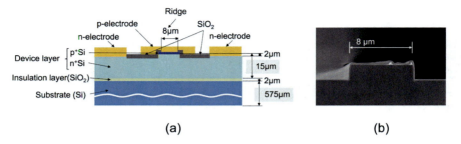

Fig. 2.8 Structure of the device. **a** A cross-sectional profile. **b** An SEM image of the end-facet of the waveguide

The visible LEDs have been fabricated by using crystalline Si that exhibit the PB phenomenon. Specifically, blue, green, and red light-emitting LEDs were fabricated by irradiating blue, green, and red light, respectively, during the DPP-assisted annealing [42]. A lateral p-n homojunction structure was developed in order to increase the efficiency of extracting the visible light from the Si crystal [43].

A variety of visible LEDs have been developed by using crystalline silicon carbide (SiC) even though it is also a well-known indirect transition-type semiconductor. They were fabricated by irradiating them with UV-violet, bluish-white, blue, and green light during the DPP-assisted annealing. The fabricated devices emitted UV-violet, bluish-white, blue, and green light, respectively [44].

2.4.2 Silicon Lasers

A simple ridge waveguide (8 μm-width and 2 μm-height) was built-in, and the cleaved facets were used as mirrors of a Fabry-Perot cavity (500 μm-cavity length). DPP-assisted annealing was carried out by injecting 1.3 μm-wavelength light into the cavity through one of the end facets. Its cross-sectional profile is schematically illustrated in Fig. 2.8a. Figure 2.8b shows a scanning electron microscopic (SEM) image of an end-facet of the waveguide.

Figure 2.9 shows the light emission spectral profiles that were acquired at room temperature (25°C) [45]. Figure 2.9a shows the profile above the threshold for lasing (the threshold current density $J = 42 \, A/cm^2$). Here, the threshold current density J_{th} was as low as $40 \, A/cm^2$. The sharp spectral peak at a wavelength of 1.40 μm represents the CW laser oscillation with a single longitudinal mode. Figure 2.9b shows the spectral profile below the threshold ($J = 38 \, A/cm^2$), in which no amplified spontaneous emission spectra are seen because of the gain depletion due to the mode competition.

By modifying the device structure above, high-power infrared laser devices were fabricated. The ridge waveguide was not built into them because further increases in the optical confinement efficiency were not expected in this waveguide as long as the

2.4 Silicon Light-Emitting Devices

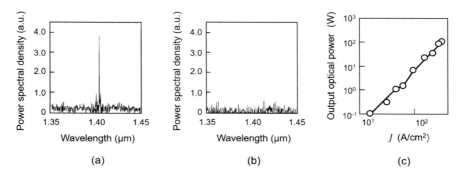

Fig. 2.9 Light emission spectra and output optical power of Si lasers. **a, b** Single-mode laser: spectral profiles above (injected current density $J = 42\,\text{A/cm}^2$) and below ($J = 38\,\text{A/cm}^2$) the threshold, respectively. The threshold current density J_{th} is $40\,\text{A/cm}^2$. **c** High-power laser: relation between J and the output optical power

device had a p-n homojunction. Instead, the cavity length was increased to 15 mm to realize high-power lasing by utilizing the very low infrared absorption of crystalline Si. As is shown in Fig. 2.9c, an output optical power higher than 100 W was obtained by further increases in the cavity length to 30 mm [46, 47].

By summarizing the experimental results of the first to the third examples above, the following phenomenon that is contrary to the common view [e] was confirmed: [Phenomenon 14] *By DPP-assisted annealing, a Si crystal works as a high-power light emitting device even though it is an indirect transition-type semiconductor.*

2.4.3 SiC Polarization Rotators

Figure 2.10a schematically explains the cross-sectional structure of the device [48]. The surface of an indirect-transition-type 4H-SiC semiconductor crystal was transformed to a p-type material by implanting Al atoms, forming a p-n junction. A ring-shaped electrode and a planar electrode were deposited on the top and bottom surfaces to form a device (Fig. 2.10b). By the DPP-assisted annealing, during which the substrate surface was irradiated with the 405 nm-wavelength light, diffusion of Al atoms was autonomously controlled to realize the optimum spatial distribution of Al atoms for light emission. As a result, the device worked as a visible LED.

A current was injected into the ring-shaped electrode for operating the fabricated device as a polarization rotator. Since a magnetic field was generated around this electrode, no external coils or bulky magnets were required to induce the magneto-optical effect. The value of the Verdet constant as high as 9.51×10^4 (rad/T.m) was obtained for linearly polarized 450 nm-wavelength laser light. This value is 10^2-times higher than those of TGG, CeF_3, and PrF_3 [49], which are the materials used

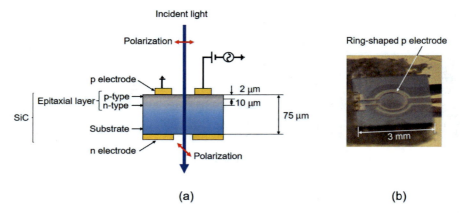

Fig. 2.10 Transmission-type polarization rotator. **a** Cross-sectional profile. **b** Photograph of the device. (By the courtesy of T. Kadowaki, Nichia Corp.)

for conventional optical isolators in the visible range, by which a gigantic magneto-optical (MO) effect was confirmed.

Figure 2.11 shows the magnified relation between the applied magnetic field H (Oe) and the magnetization M (emu/g) at 300 K obtained by subtracting the contributions of the diamagnetic component from the measured values of M. Curves A and A' are for the device after and the DPP-assisted annealing. Curve B is for before the DPP-assisted annealing. The values of M of the curves A and A' are remarkably larger than those of the curve B. The curves show hysteresis characteristics and exhibits a coercive force, as shown in the *inset* of Fig. 2.11. These characteristics indicate that the ferromagnetic characteristics were induced by the DPP-assisted annealing. It was confirmed that the saturated values of M of curves A and A' increased with increasing DPP-assisted annealing time.

By comparing these measured values, the following novel phenomenon was confirmed: [Phenomenon 15] *The semiconductor SiC crystal behaves as a ferromagnet as a result of the DPP-assisted annealing and exhibited a gigantic magneto-optical effect in the visible region.* Thus, the fabricated device exhibited a large MO effect, and this led to the large polarization-rotation of the transmitted visible light.

This behavior originated from the formation of Al atom pairs, autonomously formed as a result of the DPP-assisted annealing. This origin can be understood by referring to the following two research findings: (1) It has been found that the triplet state of the electron orbital in an Al atom pair is more stable than the singlet state [50]. (2) Two electrons with parallel spins in the triplet state induce the ferromagnetic characteristic [51].

A reflection type device has been also developed [52]. Its structure is basically the same as that of the transmission-type (Fig. 2.10a). The Verdet constant was as high as 1.83×10^6 (rad/T.m) for linearly polarized 405 nm-wavelength laser light. Furthermore, it should be pointed out that similar polarization rotators have also been

Fig. 2.11 Relations between the applied magnetic field H (Oe) and the magnetization M (emu/g) at 300 K. Curves A and A′: A SiC crystal after DPP-assisted annealing with doping Al atoms (the annealing time: 72 h). Curve B: A SiC crystal before DPP-assisted annealing. The *inset* shows the magnified relations near the origin of the graph

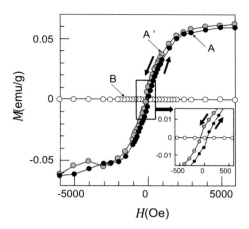

fabricated by using zinc oxide (ZnO) crystals based on the same principle as in the SiC crystals above [53].

References

1. Ohtsu, M.: Progress in dressed photon technology and the future. In: Yatsui, T. (ed.) Progress in Nanophotonics, vol. 4, pp. 1–18. Springer, Heidelberg (2018)
2. Tanaka, Y., Kobayashi, K.: Spatial localization of an optical near field in one-dimensional nanomaterial system. Physica E **40**, 297–300 (2007)
3. Sangu, S., Kobayashi, K., Ohtsu, M.: Optical near fields as photon-matter interacting systems. J. Microsc. **202**, 279–285 (2001)
4. Kawazoe, T., Ohtsu, M., Aso, S., Sawado, Y., Hosoda, Y., Yoshizawa, K., Akahane, K., Yamamoto, N., Naruse, M.: Two-dimensional array of room-temperature nanophotonic logic gates using InAs quantum dots in mesa structures. Appl. Phys. B **103**, 537–546 (2011)
5. Kawazoe, T., Tanaka, S., Ohtsu, M.: A single-photon emitter using excitation energy transfer between quantum dots. J. Nanophoton. **2**, 029502 (2008)
6. Naruse, M., Holmström, P., Kawazoe, T., Akahane, K., Yamamoto, N., Thylen, L., Ohtsu, M.: Energy dissipation in energy transfer mediated by optical near-field interactions and their interfaces with optical far-fields. Appl. Phys. Lett. **100**, 241102 (2012)
7. Naruse, M., Leibnitz, K., Peper, F., Tate, N., Nomura, W., Kawazoe, T., Murata, M., Ohtsu, M.: Autonomy in excitation transfer via optical near-field interactions and its implications for information networking. Nano Commun. Netw. **2**, 189–195 (2011)
8. Naruse, M., Tate, N., Aono, M., Ohtsu, M.: Information physics fundamentals of nanophotonics. Rep. Prog. Phys. **76**, 056401 (2013)
9. Kim, S.-J., Naruse, M., Aono, M., Ohtsu, M., Hara, M.: Decision maker based on nanoscale photo-excitation transfer. Sci. Rep. **3**, 2370 (2013)
10. Aono, M., et al.: Amoeba-inspired nanoarchitectonic computing: solving intractable computational problems using nanoscale photoexcitation transfer dynamics. Langmuir **29**, 7557–7564 (2013)
11. Tate, N., Sugiyama, H., Naruse, M., Nomura, W., Yatsui, T., Kawazoe, T., Ohtsu, M.: Quadrupole-dipole transform based on optical near-field interactions in engineered nanostructures. Opt. Express **17**, 11113–11121 (2009)

12. Tate, N., Naruse, M., Yatsui, T., Kawazoe, T., Hoga, M., Ohyagi, Y., Fukuyama, T., Kitamura, M., Ohtsu, M.: Nanophotonic code embedded in embossed hologram for hierarchical information retrieval. Opt. Express **18**, 7497–7505 (2010)
13. Kawazoe, T., Kobayashi, K., Takubo, S., Ohtsu, M.: Nonadiabatic photodissociation process using an optical near field. J. Chem. Phys. **122**, 024715 (2005)
14. Kawazoe, T., Kobayashi, K., Ohtsu, M.: Near-field optical chemical vapor deposition using Zn(acac)2 with a non-adiabatic photochemical process. Appl. Phys. B **84**, 247–251 (2006)
15. Ohtsu, M.: History, current development, and future directions of near-field optical science. Opto-Electron. Adv. **3**, 190046 (2020)
16. Lim, J., Yatsui, T., Ohtsu, M.: Observation of size-dependent resonance of near-field coupling between a deposited Zn dot and the probe apex during near-field optical chemical vapor deposition. IEICE Trans. Electron. **E88-C**, 1832–1834 (2005)
17. Polonski, V., Yamamoto, Y., Kourogi, M., Fukuda, H., Ohtsu, M.: Nanometric patterning of zinc by optical near-field photochemical vapor deposition. J. Microscopy. **194**, 545–551 (1999)
18. Yonemitsu, H., Kawazoe, T., Kobayashi, K., Ohtsu, M.: Nonadiabatic photochemical reaction and application to photolithography. J. Photoluminescence. **122–123**, 230–233 (2007)
19. Inao, Y., Nakasato, S., Kuroda, R., Ohtsu, M.: Near-field lithography as prototype nanofabrication tool. Microelectron. Eng. **84**, 705–710 (2007)
20. Kawazoe, T., Kobayashi, K., Akahane, K., Naruse, M., Yamamoto, N., Ohtsu, M.: Demonstration of nanophotonic NOT gate using near-field optically coupled quantum dots. Appl. Phys. B **84**, 243–246 (2006)
21. Kawazoe, T., Takahashi, T., Ohtsu, M.: Evaluation of the dynamic range and spatial resolution of nonadiabatic optical near-field lithography through fabrication of Fresnel zone plates. Appl. Phys. B **98**, 5–11 (2010)
22. Koike, M., Miyauchi, S., Sano, K., Imazono, T.: X-ray devices and the possibility of applying nanophotonics. In: Ohtsu, M. (ed.) Nanophotonics and Nanofabrication, pp. 179–191. Wiley-VCH, Germany (2009)
23. Hirata, K.: Realization of high-performance optical element by optical near-field etching. Proc. SPIE **7921**, 79210M (2011)
24. Yatsui, T., Nomura, W., Ohtsu, M.: Realization of ultraflat plastic film using dressed-photon-phonon-assisted selective etching of nanoscale structures. Adv. Opt. Technol. 701802 (2015)
25. Yatsui, T., Hirata, K., Tabata, Y., Miyake, Y., Akita, Y., Yoshimoto, M., Nomura, M., Kawazoe, T., Naruse, M., Ohtsu, M.: Self-organized near-field etching of the sidewalls of glass corrugations. Appl. Phys. B **103**, 527–530 (2011)
26. Teki, R., Kadaksham, A. J., House, M., Jones, J. H., Ma, A., Babu, S. V., Hariprasad, A., Dumas, P., Jenkins, R., Provine, J., Richmann, A., Stowers, J., Meyers, S., Dietze, U., Kusumoto, T., Yatsui, T., Ohtsu, M., Goodwin, F.: Alternative smoothing techniques to mitigate EUV substrate defectivity. In: Proceedings of the Society of Photo-optical Instrumentation Engineers (SPIE), vol. 8322, pp. 1–12, SPIE, San Jose, CL, USA (2012)
27. Yatsui, T., Nomura, W., Stehlin, F., Soppera, O., Naruse, M., Ohtsu, M.: Challenges in realizing ultraflat materials surfaces. Beilstein J. Nanotechnol. **4**, 875–885 (2013)
28. Nomura, W., Yatsui, T., Yanase, Y., Suzuki, K., Fujita, M., Kamata, A., Naruse, M., Ohtsu, M.: Repairing nanoscale scratched grooves on polycrystalline ceramics using optical near-field assisted sputtering. Appl. Phys. B **99**, 75–78 (2010)
29. Yatsui, T., Nomura, W., Naruse, M., Ohtsu, M.: Realization of an atomically flat surface of diamond using dressed photon-phonon etching. J. Phys. D **45**, 475302 (2012)
30. Hirschman, K.D., Tysbekov, L., Duttagupta, S.P., Fauchet, P.M.: Silicon-based visible light emitting devices integrated into microelectronic circuits. Nature **384**, 338–341 (1996). https://doi.org/10.1038/384338a0
31. Lu, Z.H., Lockwood, D.J., Baribeau, J.M.: Quantum confinement and light emission in SiO2/Si superlattices. Nature **378**, 258–260 (1995). https://doi.org/10.1038/378258a0
32. Kawazoe, T., Mueed, M.A., Ohtsu, M.: Highly efficient and broadband Si homojunction structured near-infrared light emitting diodes based on the phonon-assisted optical near-field process. Appl. Phys. B **104**, 747–754 (2011)

References

33. Ohtsu, M. (ed.): Silicon Light-Emitting Diodes and Lasers, pp. 43–51. Springer, Heidelberg (2016)
34. Yamaguchi, M., Kawazoe, T., Ohtsu, M.: Evaluating the coupling strength of electron-hole pairs and phonons in a 0.9 μm-wavelength silicon light emitting diode using dressed-photon-phonons. Appl. Phys. A. **115**, 119–125 (2013)
35. Ohtsu, M., Kawazoe, T.: Principles and practices of Si light emitting diodes using dressed photons. Adv. Mat. Lett. **10**, 860–867 (2019)
36. Tanaka, Y., Kobayashi, K.: Optical near field dressed by localized and coherent phonons. J. Microsc. **229**, 228–232 (2007)
37. Kawazoe, T., Nishioka, K., Ohtsu, M.: Polarization control of an infrared silicon light-emitting diode by dressed photons and analyses of the spatial distribution of doped boron atoms. Appl. Phys. A **121**, 1409–1415 (2015)
38. Wada, N., Kawazoe, T., Ohtsu, M.: An optical and electrical relaxation oscillator using a Si homojunction structured light emitting diode. Appl. Phys. B **108**, 25–29 (2012)
39. Sakuma, H., Ojima, I., Ohtsu, M., Kawazoe, T.: Drastic advancement in nanophotonics achieved by a new dressed photon study. J. Eur. Opt. Soc. Rapid Publ. 17–28 (2021). https://doi.org/10.1186/s41476-021-00171-w
40. Hanbury, B.R., Twiss, R.Q.: A test of new type of stellar interferometer on Sirius. Nature **178**(4541), 1046–1048 (1956)
41. Wightman, A.S.: On the localizability of quantum mechanical systems. Rev. Mod. Phys. **34**, 845 (1962)
42. Tran, M.A., Kawazoe, T., Ohtsu, M.: Fabrication of a bulk silicon p-n homojunction-structured light emitting diode showing visible electroluminescence at room temperature. Appl. Phys. A **115**, 105–111 (2014)
43. Yamaguchi, M., Kawazoe, T., Yatsui, T., Ohtsu, M.: Spectral properties of a lateral p-n homojunction-structured visible silicon light-emitting diode fabricated by dressed-photon-phonon-assisted annealing. Appl. Phys. A **121**, 1389–1394 (2015)
44. Ohtsu, M.: Silicon Light-Emitting Diodes and Lasers, pp. 83–101. Springer (2016)
45. Tanaka, H., Kawazoe, T., Ohtsu, M., Akahane, K.: Decreasing the threshold current density in Si lasers fabricated by using dressed-photons. Fluoresc. Mater. **1**, 1–7 (2015)
46. Kawazoe, T., Hashimoto, K., Sugiura, S.: High-power current-injection type silicon laser using nanophotonics. In : Abstract of the EMN Nanocrystals Meeting, pp. 9–11. Xi'an, China (Paper Number 03) (2016)
47. Ohtsu, M.: Off-Shell Application in Nanophotonics, pp. 130–136. Elsevier, Amsterdam (2021)
48. Kadowaki, T., Kawazoe, T., Ohtsu, M.: SiC transmission-type polarization rotator using a large magneto-optical effect boosted and stabilized by dressed photons. Sci. Rep. **10**, 12967 (2020)
49. Molina, P., Vasyliev, V., Villora, E.G., Shimamura, K.: CeF3 and PrF3 as UV-visible Faraday rotators. Opt. Express **19**(12), 11786 (2011)
50. Upton, T.H.: Low-lying valence electronic states of the aluminum dimer. J. Phys. Chem. **90**, 754–759 (1986)
51. Rajca, A.: Organic diradicals and polyradicals: from spin coupling to magnetism? Chem. Rev. **94**, 871–893 (1994)
52. Ohtsu, M., Kawazoe, T.: Gigantic ferromagnetic magneto-optical effect in a SiC light-emitting diode fabricated by dressed-photon-phonon-assisted annealing. Off-Shell Arch. Off-Shell: 1809R.001.v1 (2018). https://doi.org/10.14939/1809R.001.v1. https://rodrep.or.jp/en/off-shell/review_1809R.001.v1.html
53. Tate, N., Kawazoe, T., Nomura, W., Ohtsu, M.: Current-induced giant polarization rotation using ZnO single crystal doped with nitrogen ions. Sci. Rep. **5**, 12762 (2015)

Chapter 3
Preliminary Theoretical Studies and Numerical Simulations

Abstract The first part of this chapter reviews urgent theoretical analysis of the nature of the DP by using a conventional on-shell scientific method and presenting their problems. Second, novel phenomena of DP energy transfer (Table 1.1) and their experimental grounds are described. The last part reviews examples of numerical simulations for DP energy transfer by using random walk model and presents their problems. Portions of Chap. 3 have been reproduced from Ref. [7] with permission from Elsevier.

3.1 Conventional Theoretical Studies on the Dressed Photon and Their Problems

Conventional on-shell scientific methods have been employed to describe the experimentally observed nature of the DP and relevant optical phenomena. This section reviews two of these theories and their associated problems [1].

3.1.1 Creating the Dressed Photon and Coupling with Phonons

In order to describe how a DP is created on an NP, the interaction between photons and excitons in an NP has been analyzed by using conventional on-shell scientific methods [2]. Specifically, the Hamiltonian operator of the DP had to be derived for this analysis. However, since the size of the DP is much smaller than the wavelength of the incident light, a problem was that one could not define the electromagnetic mode for the DP, which is indispensable for deriving the Hamiltonian operator. In order to solve this problem, the Hamiltonian operator was temporarily expressed by the sum of the operators of an infinite number of electromagnetic modes of a free photon, based on conventional Fourier transform analysis. Furthermore, the number of energy levels of the exciton was assumed to be infinite. By diagonalizing this temporary Hamiltonian operator, the creation and annihilation operators of a quasi-

© The Author(s), under exclusive license to Springer Nature Switzerland AG 2025
M. Ohtsu and H. Sakuma, *Dressed Photons to Revolutionize Modern Physics*,
Nano-Optics and Nanophotonics, https://doi.org/10.1007/978-3-031-77944-2_3

particle participating in the interaction were derived. Since these operators are given by the sum of the operators for a photon and an exciton, they represent a novel quantum field that is created as a result of coupling between the photon and the exciton. This field is a photon that dresses the exciton energy, which is the origin of the name DP.

In the case where a crystalline NP is irradiated with light, a DP is created and localized at the i-th atom in the crystal. Then, the DP hops to the adjacent atoms and excites phonons. Finally, the DP couples with these phonons. The creation operator of the novel quasi-particle, created as a result of the DP-phonon coupling that is localized at the i-th atom was derived [3]. Since this operator is given by the product of the operators of the DP and the displacement operator function of phonons, it indicates that the DP excites multimode coherent phonons and couples with them. That is, the quantum field created as a result of this excitation and coupling is a new type of DP that *dresses* the phonon energy. This is the dressed-photon-phonon (DPP) (refer to Sect. 2.2). As a result of this coupling, the DP energy is expressed as $E_{DP} = h\nu_{in} + E_{exciton} + E_{phonon}$, which is larger than the incident photon energy ($h\nu_{in}$), where $E_{exciton}$ and E_{phonon} are the energies of the exciton and phonon, respectively.

3.1.2 Localization of the Dressed Photon

In order to describe the spatially localized nature of the DP, two NPs placed in close proximity to each other were assumed. The DP energy was bidirectionally transferred between them. It was also assumed that the microscopic system (composed of the DP and the two NPs (NP1 and NP2)) was buried in a macroscopic system (composed of a macroscopic host material and macroscopic incident light). For these two systems, the method of renormalization was employed so that the contribution from the macroscopic system to the DP energy transfer was expressed as the effective interaction energy between the two NPs. As a result, the magnitude of the effective interaction energy localized on NPi ($i = 1, 2$) was given by the Yukawa-type function $V_i(r) \propto \exp(-r/a_i)/r$, where r is the distance measured from the center of NPi [4]. Since the spatial extent of this function is given by the size a_i of the NPi, it was confirmed that the spatial extent of the localized DP was equivalent to the size of the NP.

It should be noted that this subwavelength-sized localization of the DP violates the long-wavelength approximation that has been employed to calculate the electric-dipole interaction probability for conventional light-electron interactions (on-shell science). Due to this violation, a conventional dipole-forbidden transition became an allowed transition in the case of the DP-electron interaction [Phenomenon 5].

3.1.3 Theoretical Problems and the Road to a Solution

It should be pointed out that the conventional theoretical studies reviewed in Sect. 3.1.1 and 3.1.2 have several problems. In the case of Sect. 3.1.1, the problem was that the electromagnetic modes could not be defined for the DP due to its subwavelength size. To solve this problem, an infinite number of electromagnetic modes were summed, and a temporary solution was derived. However, such a summation of the modes could not describe the interaction between photons and excitons in an NP. That is, the DP (off-shell quantum field) could not be described even when infinite numbers of lines and curves in Fig. 2.1 (on-shell fields) were superposed. In the case of Sect. 3.1.2, a macroscopic system was required to describe the localization of the DP created on the two NPs. Curiously, the DP on the two NPs could not be described when they were installed in a vacuum. The problems above indicate that the development of genuine off-shell scientific theories is required.

Even under such a situation, experimental studies on the DP have been extensively carried out in the last two decades, and their applications have resulted in novel generic technologies (Sects. 2.3 and 2.4) [5]. Some of them have already been put into practical use. The development of theoretical studies of off-shell science is required to promote further developments in the application of these technologies.

3.2 Spatial Evolution of DP Energy Transfer

This section focuses on Phenomena 4–7, 9, and 11–14 in Table 1.1 again. They require novel off-shell scientific theories for realizing future progress in novel technologies reviewed in Sects. 2.3 and 2.4 [6, 7].

3.2.1 Size-Dependent Resonance and Hierarchy

This subsection starts by reviewing the efficiency of the DP energy transfer between the two spherical NPs (NP1 and NP2, with radii a_1 and a_2, respectively: Fig. 3.1a). Section 3.1.2 presented the Yukawa function that represented the magnitude of the interaction energy between the two NPs mediated by a DP [8]. As a result of the interaction, propagating light was created from the NPs and could be measured by a conventional photodetector installed in the far field. The intensity of this light is shown in Fig. 3.1b, that was temporarily derived by using a conventional on-shell scientific theory [9]. The curve in this figure shows that the efficiency resonantly takes the maximum when $a_1 = a_2$. This feature has been called size-dependent resonance [Phenomenon 4].

The discussions on hierarchy can be started based on the size-dependent resonance [Phenomenon 4]. Here, it is assumed that two spherical NPs (NP1 and NP2 with radii

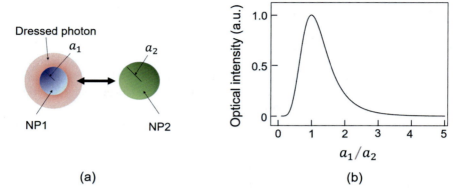

Fig. 3.1 Size-dependent resonance. **a** Two spherical NPs and their radii. **b** Relation between the radius and the measurable intensity of the propagating light. a_2 was fixed to 20 nm. The surface-to-surface separation between the two NPs is 1 nm

a_1 and a_2, respectively) are installed in close proximity to each other (Fig. 3.2). It is also assumed that two more NPs (NP1′ and NP2′, with radii a_1' ($> a_1$) and a_2' ($> a_2$), respectively) are installed in proximity to NP1 and NP2. The size-dependent resonance realizes efficient energy transfer of the DP on NP1 to NP2 when $a_1 = a_2$. Although the energy on NP1′ is also efficiently transferred to NP2′ when $a_1' = a_2'$, the efficiency of the DP energy transfer to NP2 is low due to the size difference ($a_2 \neq a_2'$). The efficiency of the transfer from NP1 to NP2′ is also low. That is, the channels of the DP energy transfer between the different-sized pairs do not exhibit any crosstalk. This feature of DP energy transfer without any crosstalk is called hierarchy [Phenomenon 7]. It means that different energy transfers occur independently for different material sizes.

For further discussions on hierarchy, one should consider the size of the material. In the case of a spherical NP, its size is represented by its radius. However, even though it is recognized as a sphere when it is viewed in the far field, its surface often has roughness when it is viewed in the near field. That is, the recognized shape and size depend on the separation between the NP and the observer. The hierarchy is related to these separation-dependencies. If the surface of the above-mentioned spherical NP is divided into small parts, and they are approximated as spheres whose radii are equivalent to the size of the roughness, discussions equivalent to those of the original spherical NP can be made. The concept of hierarchy is established by assuming that the spatial features of the divided parts are equivalent to those of the original spherical NP.

However, this division cannot be repeated infinitely. The minimum size of the NP to be divided obviously corresponds to the size of an atom, for which the discussions of hierarchy above are valid. On the other hand, experimental studies have estimated that the maximum size of the DP was 40–70 nm [Phenomenon 9], which corresponds to the maximum size of the NP for which discussions of hierarchy are valid.

3.2 Spatial Evolution of DP Energy Transfer

Fig. 3.2 Schematic explanation of hierarchy

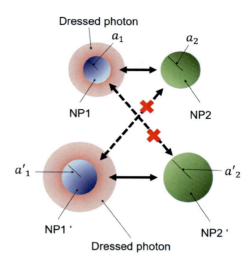

3.2.2 Autonomy

Experiments have observed unique characteristics of the DP energy transfer among NPs. From these characteristics, it appears as if DP energy transfer occurs of its own will, and thus this behavior has been called autonomy [Phenomenon 6]. One of its origins has been attributed to the size-dependent momentum resonance and hierarchy. These characteristics offer a curious resemblance to the behavior of an organism. Two examples {1} and {2} are reviewed below.

{1} The DP indicates its existence to the macroscopic system most effectively. The experimental ground for this indication is a DP energy transmitter that is composed of an array of NPs (Fig. 3.3): The DP in the nanometer-sized system autonomously selects the route for transferring its energy so as to maximize the power of the generated propagating light (the output signal) [10]. Such the autonomous DP energy transfer has been observed and found that the efficiency of the photocurrent generation from a photodiode took the maximum when the ratio of the numbers of small and large NPs are 4:1 [11]. Furthermore, it was found that the efficiency is higher when some cannels of DP energy transfer are degraded [12]. The energy transfer process described above is similar to that in a photosynthetic bacteria [13].

{2} The DP indicates that it minimizes the magnitude of the energy dissipation of the macroscopic system by removing the DP energy from the nanometer-sized system most effectively. The experimental grounds for this indication are:

{2-1} [Smoothing of a material surface (Sect. 2.3.2)] The DPP autonomously annihilates so as to minimize the energy dissipation of incident light during the process of etching a bump on a material surface.

{2-2} [Photovoltaic device, fabricated and operated by DPP] [14] (Fig. 3.4) The DPP autonomously modifies the spatial distribution of silver (Ag) particles so as to

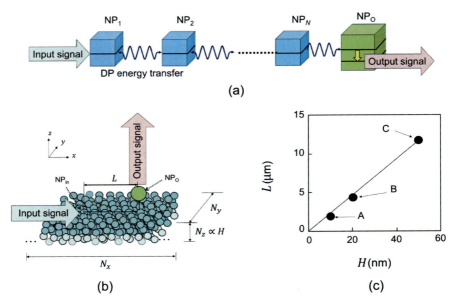

Fig. 3.3 DP energy transmitter. **a** Schematic explanation of operation (refer to Fig. 3.8). **b** Spherical CdSe NPs (average diameters were 2.8 nm and 4.1 nm for NP_1–NP_N and NP_O, respectively) were dispersed on a SiO_2 substrate. The average separation between the adjacent NPs was arranged to be close to 7.3 nm in order allow efficient DP energy transfer. The thickness of the NP layers, H was fixed to 10, 20, and 50 nm, which is proportional to the number of rows N_z of NP_1–NP_N along the z-axis. These devices are denoted by A, B, and C, respectively. **c** Measured dependence of the energy transfer length L on the thickness H of the small NP layers

maximize the output photocurrent when the input light has the same photon energy as that of the light irradiated during the device fabrication (photon breeding).

{2-3} [Silicon light-emitting devices (Sect. 2.4)] The DPP autonomously modifies the spatial distribution of boron (B) atoms so as to maximize the emitted light power whose photon energy is equivalent to that of light irradiated during the device fabrication (photon breeding (PB)) [Phenomena 11–14]. Furthermore, the B atoms form pairs, and these pairs induce photon breeding. This induction is analogous to the induction of the self-duplicating function originating from the pair of helices in DNA. The PB indicates that the light emitted from the device is a replica of the light irradiated on the crystal during the device fabrication. That is, the emitted light is self-duplicated by the irradiated light, which suggests that the behavior of the DP is analogous to that of organisms. It should be pointed out that an infrared Si-photodiodes, fabricated by the DPP-assisted annealing, also exhibits the PB in its spectral photosensitivity (Fig. 3.5) [15].

It should be pointed out that the autonomy can be found not only in the phenomena originated from DP but also in a variety of biological and chemical phenomena as long as electromagnetic fields or photon are engaged in. However, their origins have never been studied. It is expected that off-shell science can promote these studies.

3.2 Spatial Evolution of DP Energy Transfer

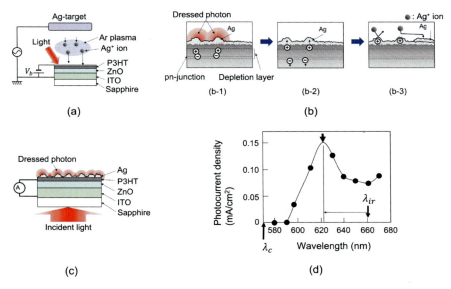

Fig. 3.4 Photovoltaic devices using organic molecules of poly(3-hexylthiophene) (P3HT). **a** Deposition of Ag particles by rf-sputtering under light irradiation (the wavelength λ_{ir} is 660 nm (> the cutoff wavelength λ_c (= 570 nm) of the P3HT). **b** Principle of controlling the amount of Ag particles that flow into and out of the Ag film surface. (b-1), (b-2), and (b-3) represent the creation of electron-hole pairs by the DP, charging of the Ag film, and autonomous control of the Ag particle deposition, respectively. **c** Operation of the fabricated photovoltaic device. **d** Relation between the incident light wavelength and the photocurrent density. The thick downward arrow at the peak of the curve indicates the photon breeding (PB), as was demonstrated in Sect. 2.4. The difference between λ_{ir} and the wavelength at this peak originated from the DC Stark effect induced by the reverse bias voltage V_b applied during the fabrication process (refer to the left-pointing arrow)

As an urgent theoretical analysis of the experimental results that originated from the autonomy, a random walk model has been used for numerical simulation techniques relying on conventional statistical mechanics and complex-system science. The results are reviewed in Sect. 3.3.

3.2.3 Energy Disturbance

Three kinds of NPs [1]–[3] were used as specimens to demonstrate the DP energy disturbance: [1] The first specimen contained GaN NPs[16] that were buried immediately under the surface of an AlN substrate. Their diameters were 50–70 nm, and their heights were 7–10 nm. Curve A in Fig. 3.6 shows the photoluminescence (PL) spectral profile acquired by using a conventional microscope (on-shell science).

Since the peak energies of the narrow PL spectra from the GaN NPs depended on their scattering sizes, the curve A corresponded to the envelope of these scattered

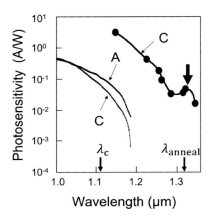

Fig. 3.5 Photosensitivity of the infrared Si-photodiodes. λ_c is the cutoff-wavelength determined by E_g. λ_{anneal} is the wavelength of the light irradiated during the DPP-assisted annealing. Curves A shows the values obtained with a Si-PD fabricated with the DPP-assisted annealing without the forward current density. Curve B is the values with the forward current density of 10.0 A/cm^2. Curve C shows the values obtained with a Si-PIN photodiode (Hamamatsu Photonics, S3590) used as a reference. The downward thick arrow indicates the photon breeding (PB)

Fig. 3.6 PL spectral profiles from GaN NPs. Curve A and B are the spectra obtained using methods of on-shell science and off-shell science respectively

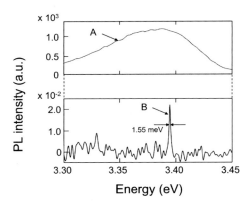

narrow spectra from a large number of the GaN NPs, and thus, its width was very large. On the other hand, in the case where the fiber probe was used (off-shell science), it was expected that a few narrow spectra emitted from a few NPs located under the probe tip would be resolved.

However, as shown by the curve B, only an extremely narrow PL spectrum that originated from a single NP located exactly under the probe tip was acquired. This was because the DP energy was preferentially transferred from this NP to the probe tip most efficiently. This indicates that the DP energy transfers from other NPs, located slightly away from the probe tip, were suppressed, indicating that the linear relation between the cause and effect of the DP energy measurement was lost, thus suggesting energy disturbance of the excitons and the DP.

3.2 Spatial Evolution of DP Energy Transfer 35

Fig. 3.7 Temporal variation of the light intensity emitted from the electric-dipole forbidden energy level. Closed circles represent the measured value. Curves A and B are the exponential functions fitted to them. Curve A: The probe-specimen distance was 5 nm. The emission lifetime estimated from the fitted curve was 305 ps. Curve B: The probe-specimen distance was <5 nm. The emission lifetime estimated from the fitted curve was 260 ps

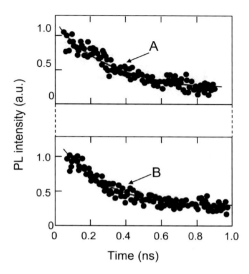

[2] The second specimen contained nanometer-sized rings (NR) of GaAs that were buried immediately under the surface of an AlGaAs substrate [17]. It should be pointed out that the lowest excited energy level of the exciton in the NR is electric-dipole forbidden. PL emission spectra from these NRs were acquired by using the fiber probe. Here, the NRs were irradiated with a short optical pulse in order to measure the temporal variation of the emitted light intensity. From these measurements, light emission from the electric-dipole forbidden energy level was clearly seen at a temperature of 7 K; such an effect has never been seen using the methods of on-shell science. This is evidence of violation of the long-wavelength approximation [Phenomenon 5].

Curve A in Fig. 3.7 shows the temporal variation of the light intensity emitted from the electric-dipole forbidden level, which demonstrated an emission lifetime of 305 ps when the probe-specimen distance was 5 nm. Such a short lifetime indicates that the fiber probe disturbed the exciton energy and also the DP to trigger light emission from the forbidden state. As shown by the curve B, decreases of the fiber-specimen distance decreased the emission lifetime to 260 ps. This decrease in the lifetime indicates that the energy disturbance was enhanced by decreasing the probe-specimen distance.

[3] The last specimen contained nanometer-sized cubic NPs (NP1 and NP2 in Fig. 3.8) of CuCl that were grown in an NaCl host crystal [18]. The quantum numbers (1,1,1) in Fig. 3.8 represent the lowest excited energy level of the exciton. By irradiating light that was resonant with this energy level in NP1, an exciton was excited, creating the DP. A larger cubic NP (NP2) was installed in proximity to NP1. It played the role of a fiber probe tip for measuring the DP. Here, it should be noted that the second excited energy level (2,1,1) in NP2 was electric-dipole forbidden. However, in the case where it was resonant with the (1,1,1) level of NP1, the DP transferred

Fig. 3.8 DP energy transfer between cubic NPs, subsequent exciton relaxation, and light emission. Two small closed circles represent the excited excitons

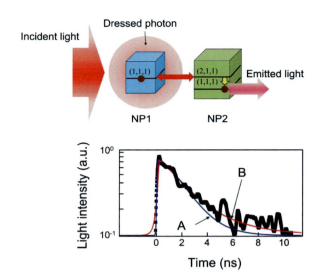

Fig. 3.9 Temporal evolution of the light intensity emitted from the CuCl NP2 for $0 \leq t \leq 10$ ns. Closed squares represent the measured value. Curves A and B are the exponential functions $\exp(-t/\tau_{f1})$ and $\exp(-\sqrt{t/\tau_{f2}})$, respectively, fitted to them

from NP1 to NP2, turning the forbidden transition to an allowed transition due to the violation of the long-wavelength approximation [Phenomenon 5]. Thus, the exciton was excited to the (2,1,1) level. This exciton subsequently relaxed to the lower level (1,1,1) to emit light that could be measured by a conventional photodetector. This light emission is evidence of disturbance of the electric-dipole forbidden energy level of the exciton and also the DP energy.

For studying the radiative relaxation process, closed squares in Fig. 3.9 show the measured temporal evolution of the PL from the lower level (1,1,1) of NP2 in the time span as long as 0–10 ns [19]. The curve A represents the temporal evolution expressed as $\exp(-t/\tau_{f1})$, where τ_{f1} is the fall time, depending on the magnitude of the transferred DP energy. This curve agrees with the closed squares only for an initial stage as short as $0 \leq t \leq 2$ ns. On the other hand, the curve B represents the temporal evolution expressed as $\exp(-\sqrt{t/\tau_{f2}})$, where the fall time τ_{f2} is the radiative relaxation rate from the (1,1,1) state in NP2. This agrees with the closed squares for a wide range of time periods up to 10 ns. The component expressed as $\exp(-\sqrt{t/\tau_{f2}})$ can be suppressed by decreasing the device temperature.

The temporal evolutions shown in Fig. 3.9 have two features: {1} The temporal evolution expressed as $\exp(-t/\tau_{f1})$ originated from the DP energy transfer between NPs. The rise time τ_r also originates from this transfer. {2} The temporal evolution expressed as $\exp(-\sqrt{t/\tau_{f2}})$ originated from the radiative relaxation in each NP.

Features {1} and {2} above represent unique phenomena which are different from each other. The former is exactly the novel off-shell scientific phenomenon [20]. The latter is no more than a conventional on-shell phenomenon. The fact that these temporal evolutions are respectively expressed as $\exp(-t/\tau_{f1})$ and $\exp(-\sqrt{t/\tau_{f2}})$ suggests that they correspond to the quantum walk [21] and the random walk relaxation processes, respectively. As for the former process, it should not be considered as a mere random walk because its energy transportation is linearly dependent on time, not on the square root of time.

3.3 Numerical Simulations and Their Problems

In order to describe the autonomy [Phenomenon 6] observed in DP energy transfer and its measurement process, numerical simulations have been carried out by temporarily using a random walk model relying on statistical mechanics and complex-systems science (on-shell science). This section reviews the results of these simulations and presents the problems associated with them [22].

The origin of the PB effect in the Si-LED [Phenomenon 13], reviewed in Sect. 2.4, was that the spatial distribution of B atoms played the role of genes. That is, this distribution bore the generic information and autonomously varied depending on the photon energy and polarization of the light irradiated during the DPP-assisted annealing, eventually reaching a stationary state [Phenomena 11 and 12]. The spatial distribution of B atoms and the characteristics of the emitted light were simulated by a numerical simulation using a nonequilibrium statistical mechanical model [23].

The results obtained by the numerical simulation were as follows: <<Fabrication>> The regions A and B in Fig. 3.10 show the temporal variation of the simulated power of the light emitted during the DPP-assisted annealing [24]. In region A, the power increased immediately after the DPP-assisted annealing started. Then it showed relaxation oscillation, such that the amplitude of the oscillation decreased with time. Subsequently, in region B, the power reached the stationary state after a certain time and showed a relatively small fluctuation.

<<Operation>> Region C in Fig. 3.10 shows the power emitted during the device operation. Its value was smaller than those in regions A and B because of the smaller injection current than those injected for the DPP-assisted annealing. However, it was

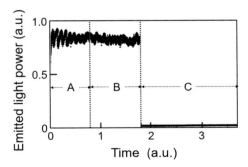

Fig. 3.10 Output power from the device. The regions A and B represent the state of transition and a stationary state during the device fabrication. The region C is the output power emitted during the device operation

stable over time. This was because the spatial distribution of the B atom-pairs in the Si crystal was fixed, and random walks of the B atoms were suppressed.

The PB effect with respect to photon spin has been also experimentally observed, as was briefly pointed out in Sect. 2.4. In this effect, the polarization of the emitted light was equivalent to that of the light irradiated during the DPP-assisted annealing. Numerical simulations above reproduced this effect by counting the number of created photon pairs that were orthogonally polarized.

Although the results of the numerical simulations above were consistent with several experimental results, there are problems that remain to be solved to explore future directions:

[1] It was not straightforward to describe the details of the PB effect when using the conventional nonequilibrium statistical mechanical model, which relied on the temperature-dependent thermal diffusion.

[2] In the conventional simulations, the numerical values were carefully selected in order to fit the results to the experimental results. However, the essential problem is how to identify the origins of the DP creation and the autonomous energy transfer of the DP. That is, we need to answer the questions "What is the DP?" and "What is the intrinsic nature of the DP?" The answers to these questions have not yet been given by these simulations.

[3] The answer to the question "What is the nature of the interaction between NPs via the DP?" has not yet been given either. The absence of an answer originates in the intrinsic nature of on-shell science. One example was that even if infinite numbers of electromagnetic modes in the on-shell region were superposed, this superposition cannot represent the off-shell electromagnetic field that plays an essential role in the interaction.

Several discussions have been made, and it has been suggested that it was advantageous to employ a quantum walk (QW) model in the numerical simulations for solving problems [1] and [2] [25]. This simulation is expected to give the answers for describing the off-shell scientific phenomena by developing a QW model with an infinite number of degrees of freedom. The line-graph method could be advantageously used for this description [26].

References

1. Ohtsu, M.: Dressed photon phenomena that demand off-shell scientific theories. In: Off-shell Archive (November, 2019) OffShell: 1911R.001.v1. https://rodrep.or.jp/en/off-shell/review_1911R.001.v1.html
2. Ohtsu, M.: Dressed Photons, pp. 11–18. Springer, Heidelberg (2014)
3. Kobayashi, K., Tanaka, Y., Kawazoe, T., Ohtsu, M.: Localized photon inducing phonons' degrees of freedom. In: Ohtsu, M. (ed.) Progress in Nano-Electro-Optics VI, pp. 41–66. Springer, Heidelberg (2008)
4. Kobayashi, K., Sangu, S., Ito, H., Ohtsu, M.: Near-field optical potential for a neutral atom. Phys. Rev. A **63**, 013806 (2001)
5. Ohtsu, M.: Dressed Photons, pp. 89–214. Springer (2014)

References

6. Ohtsu, M.: Indications from dressed photons to macroscopic systems based on hierarchy and autonomy, Off-shell Archive, Offshell: 1906R.001.v1. (2019). https://rodrep.or.jp/en/off-shell/review_1906R.001.v1.html
7. Ohtsu, M.: Off-Shell Application in Nanophotonics, pp. 23–30. Elsevier, Amsterdam (2021)
8. Kobayashi, K., Ohtsu, M.: Quantum theoretical approach to a near-field optical system. J. Microsc. **194**, 249–254 (1999)
9. Sangu, S., Kobayashi, K., Ohtsu, M.: Optical near fields as photon-matter interacting systems. J. Microsc. **202**, 279–285 (2001)
10. Nomura, W., Yatsui, T., Kawazoe, T., Naruse, M., Ohtsu, M.: Structural dependency of optical excitation transfer via optical near-field interactions between semiconductor quantum dots. Appl. Phys. B **100**, 181–187 (2010)
11. Naruse, M., Kawazoe, T., Nomura, W., Ohtsu, M.: Optimum mixture of randomly dispersed quantum dots for optical excitation transfer via optical near-field interactions. Phys. Rev. B **80**, 125325 (2009)
12. Naruse, M., Leibnitz, K., Peper, F., Tate, N., Nomura, W., Kawazoe, T., Murata, M., Ohtsu, M.: Autonomy in excitation transfer via optical near-field interactions and its implications for information networking. Nano Commun. Netw. **2**, 189–195 (2011)
13. Imahori, H., Multiporphyrin, G.: Arrays as artificial light-harvesting antennas. J. Phys. Chem. B **108**, 6130–6143 (2004)
14. Yukutake, S., Kawazoe, T., Yatsui, T., Nomura, W., Kitamura, K., Ohtsu, M.: Selective photocurrent generation in the transparent wavelength range of a semiconductor photovoltaic device using a phonon-assisted optical near-field process. Appl. Phys. B **99**, 415–422 (2010)
15. Tanaka, H., Kawazoe, T., Ohtsu, M.: Increasing Si photodetector photosensitivity in near-infrared region and manifestation of optical amplification by dressed photons. Appl. Phys. B **108**, 51–56 (2012)
16. Neogi, A., Gorman, B.P., Morkoç, H., Kawazoe, T., Ohtsu, M.: Near-field optical spectroscopy and microscopy of self-assembled GaN/AlN nanostructures. Appl. Phys. Lett. **86**, 043103 (2005)
17. Yatsui, T., Nomura, W., Mano, T., Miyazaki, H.T., Sakoda, K., Kawazoe, T., Ohtsu, M.: Emission from a dipole-forbidden energy state in a GaAs quantum-ring induced by dressed photon. Appl. Phys. A **115**(1), 1–4 (2014)
18. Kawazoe, T., Kobayashi, K., Sangu, S., Ohtsu, M.: Demonstration of a nanophotonic switching operation by optical near-field energy transfer. Appl. Phys. Lett. **82**, 2957–2959 (2003)
19. Ohtsu, M., Kawazoe, T., Saigo, H.: Spatial and Temporal Evolutions of Dressed Photon Energy Transfer. Off-shell Archive, Offshell:1710R.001.v1. (2017). https://rodrep.or.jp/en/off-shell/review_1710R.001.v1.html
20. Ohtsu, M.: New Routes to Studying the Dressed Photon. Off-shell Archive, OffShell: 1709R.001.v1. (2017). https://rodrep.or.jp/en/off-shell/review_1709R.001.v1.html
21. Konno, N.: Quantum walk. In: Franz, U., Schürmann, M.: (eds.) Quantum Potential Theory, pp. 309–452. Springer, Heidelberg (2008)
22. Ohtsu, M.: "The present and future of numerical simulation techniques for off-shell science," Off-shell Archive (March, 2020) OffShell: 2003R.001.v1. (March, 2020). https://rodrep.or.jp/en/off-shell/review_2003R.001.v1.html
23. Katori, M., Kobayashi, H.: Nonequilibrium statistical mechanical models for photon breeding processes assisted by dressed-photon-phonons. In: Ohtsu, M., Yatsui, T. (eds.) Progress Nanophotonics 4, pp. 19–55. Springer, Heidelberg (2018)
24. Ohtsu, M., Katori, M.: Complex system of dressed photons and applications. Rev. Laser Eng. **45**, 139–143 (2017)
25. Saigo, H.: Quantum probability for dressed photons: the arcsine law in nanophotonics. In: Yatsui, T. (ed.) Progress in Nanophotonics 5. Springer Nature, Switzerland AG (2018)
26. Konno, N., Portugal, R., Sato, I., Segawa, E.: Partition-based discrete-time quantum walks. Quant. Inf. Process. **17**(100) (2018)

Chapter 4
A Quantum Walk Model for the Dressed Photon Energy Transfer

Abstract This chapter reviews the numerical simulation of the DP energy transfer by an off-shell scientific method. It starts by explaining the reason why the quantum walk model is used for this simulation. After equations used for this model are presented, three simulated examples are demonstrated. They are the DPP creation probability on the tip of a fiber probe, the spatial distribution of the DPP that is confined by the B atom-pair in a Si crystal, and the photon breeding with respect to the photon spin. It is pointed out that these results agree well with the experimental results. Portions of Chap. 4 have been reproduced from Ref. [4] with permission from Elsevier.

4.1 A Quantum Walk Model

Section 3.3 reviewed that the random walk (RW) model has been temporarily used for numerical simulation on DP energy transfer. In these calculations, the numerical values used for calculations were carefully selected in order to fit the results to the experimental results. Although the essential question is how to identify the origins of the autonomous DP energy transfer, the RW model has not yet answered this question. One reason is that the RW model has never dealt with the light-matter interaction process. On the other hand, the theories reviewed in Chap. 5 deals with this process and draws a clear picture of the DP creation. Prior to this review, this chapter discusses the numerical simulation for analyzing the spatial-temporal evolution of the DPP energy transfer by phenomenologically taking the interaction process into account. A quantum walk (QW) model is used for this simulation.

The QW model is effective for describing the intrinsic features of the DPP energy transfer. This effectiveness is attributed to two origins:

(1) Non-commutativity: The QW model is based on non-commutative algebra using vectors and matrices [1]. On the other hand, the DP is a quantum field that mediates the non-commutative interaction between nanometer-sized particles (NP) (refer to Chaps. 5–10). Thus, one origin of the effectiveness above is attributed to the non-commutative aspect that is common to the QW and DP.

© The Author(s), under exclusive license to Springer Nature Switzerland AG 2025
M. Ohtsu and H. Sakuma, *Dressed Photons to Revolutionize Modern Physics*,
Nano-Optics and Nanophotonics, https://doi.org/10.1007/978-3-031-77944-2_4

41

42   4 A Quantum Walk Model for the Dressed Photon Energy Transfer

(2) Localization: The QW model treats the energy transfer from one site to its nearest-neighbor site, both of which represent local positions in the QW lattice. On the other hand, since the DP is a localized field (refer to Chaps. 5–10), its quantum mechanical position operator can be defined. Thus, if the NP is considered as the site of the QW model, the position of the DP is identified with that of this site. This implies that the other origin of the effectiveness above is the localization aspect that is common to the QW and DP.

Section 4.2 presents equations used for numerical simulation by the QW model. Section 4.3 reviews the calculated results of the DPP creation probability on the tip of a fiber probe. This probability corresponds to the throughput of creating the DPP on the tip of a fiber probe that has never been derived by the conventional wave optical model or the RW model. Section 4.4 reviews the calculated results of the DPP that is confined by the B atom-pair in a Si crystal. It has never been derived by the RW model either. Section 4.5 describes the features of photon breeding with respect to the photon spin. Not only the successful derivations of these result, it should be pointed out that the results in Sects. 4.3–4.5 agree well with the experimental results.

4.2 Equations for the Two-Dimensional Quantum Walk Model

The energy of the created DP transfers to the adjacent NP via a process called DP hopping (hopping energy J). In the case where multiple NPs are the atoms in a crystal lattice, the DP excites a lattice vibration during the hopping, resulting in the creation of phonons. Subsequently, the DP couples with these phonons (coupling energy χ) to create a DPP quantum field (Sect. 2.2).

The square lattice in Fig. 4.1a is used for a two-dimensional QW model [2, 3]. By irradiating light (an input signal that propagates along the $+y$-axis) to the lower side of this lattice, DPs are created at the sites in the lattice and travel in the upper-right or lower-left directions. These directions correspond to directions parallel and antiparallel (along the $+y$ and $-y$ axes, respectively) to the direction of the irradiated light propagation. Since the QW model deals with the DP hopping to the nearest-neighbor site, these directions are represented by red and blue bent arrows, respectively, in Fig. 4.1b, c. The phonon is represented by a green closed loop because it does not hop due to its nonlocalized nature.

Since the DPP is created as a result of coupling between two counter-travelling DPs and a phonon, a three-dimensional vector

$$\Psi_{t,(x,y)} = \begin{bmatrix} y_{DP+} \\ y_{DP-} \\ y_{Phonon} \end{bmatrix}_{t,(x,y)} \tag{4.1}$$

4.2 Equations for the Two-Dimensional Quantum Walk Model

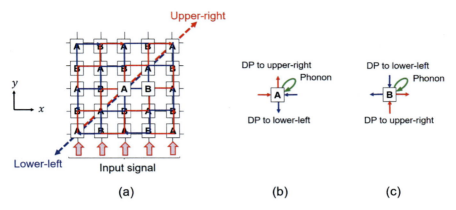

Fig. 4.1 Two-dimensional square lattice. **a** DPs that travel in the upper-right and lower-left directions. They are represented by bent red and blue arrows, respectively. **b, c** The magnified figures at sites A and B in (a), respectively. The green loop represents a phonon

is used to represent its creation probability amplitude, where [ ] is the vector at time t and at the position of the lattice site (x, y), y_{DP+} and y_{DP-} are the creation probability amplitudes of the DPs that travel by repeating the hopping in the upper-right and lower-left directions, respectively, and y_{Phonon} is that of the phonon.

First, the spatial-temporal evolution equation for the DPP, hopping out from the site A in Fig. 4.1b, is given by

$$\Psi_{t,(x,y)\updownarrow} = P_+ \Psi_{t-1,(x-1,y)\rightarrow} + P_- \Psi_{t-1,(x+1,y)\leftarrow} + P_0 \Psi_{t-1,(x,y)}. \tag{4.2}$$

The vector $\Psi_{t,(x,y)\updownarrow}$ on the left-hand side is composed of two DPs ($y_{DP+\uparrow}$ and $y_{DP-\downarrow}$: hopping out along the $\pm y$ axes) and a phonon (y_{Phonon}) at time t. The right-hand side is composed of two DPs ($y_{DP+\rightarrow}$ and $y_{DP-\leftarrow}$: hopping into site A along the $\pm x$ axes) and a phonon (y_{Phonon}). The three matrices on the right-hand side are

$$P_+ = \begin{bmatrix} \epsilon_+ & J & \chi \\ 0 & 0 & 0 \\ 0 & 0 & 0 \end{bmatrix}, \quad P_- = \begin{bmatrix} 0 & J & \chi \\ J & \epsilon_- & \chi \\ 0 & 0 & 0 \end{bmatrix}, \quad P_0 = \begin{bmatrix} 0 & 0 & 0 \\ 0 & 0 & 0 \\ \chi & \chi & \epsilon_0 \end{bmatrix}.$$
$$(3a, b, c)$$

Diagonal elements ϵ_+ and ϵ_- are the eigen-energies of the DPs (y_{DP+} and y_{DP-}), respectively, and ϵ_0 is that of the phonon. Off-diagonal elements J and χ represent the DP hopping energy and the DP-phonon coupling energy, respectively.

Second, the spatial-temporal evolution equation for the DPP, hopping out from the site B in Fig. 4.1c, is given by

$$\Psi_{t,(x,y)\leftrightarrow} = P_+ \Psi_{t-1,(x,y-1)\uparrow} + P_- \Psi_{t-1,(x,y+1)\downarrow} + P_0 \Psi_{t-1,(x,y)}. \tag{4.4}$$

The vector $\boldsymbol{\Psi}_{t,(x,y)\leftrightarrow}$ on the left-hand side is composed of two DPs ($y_{DP+\rightarrow}$ and $y_{DP-\leftarrow}$: hopping out along the $\pm x$ axes) and a phonon (y_{Phonon}) at time t. The right-hand side is composed of two DPs ($y_{DP+\uparrow}$ and $y_{DP-\downarrow}$: hopping into the site B along the $\pm y$ axes) and a phonon (y_{Phonon}). The three matrices of $3(a-c)$ are used also in (4.4).

By using vectors

$$\boldsymbol{\Psi}_{t,(x,y)\leftrightarrow} := \begin{bmatrix} y_{DP+\rightarrow} \\ y_{DP-\leftarrow} \\ y_{Phonon} \end{bmatrix}_{t,(x,y)} \quad , \quad \boldsymbol{\Psi}_{t,(x,y)\updownarrow} := \begin{bmatrix} y_{DP+\uparrow} \\ y_{DP-\downarrow} \\ y_{Phonon} \end{bmatrix}_{t,(x,y)} \quad , \quad (5a, b)$$

(4.2) and (4.4) are lumped together and represented by

$$\boldsymbol{\Psi}_{t,(x,y)} = \begin{bmatrix} \boldsymbol{\Psi}_{t,(x,y)\leftrightarrow} \\ \boldsymbol{\Psi}_{t,(x,y)\updownarrow} \end{bmatrix} = \begin{bmatrix} [0] & U \\ U & [0] \end{bmatrix} \begin{bmatrix} \boldsymbol{\Psi}_{t-1,(x,y)\leftrightarrow} \\ \boldsymbol{\Psi}_{t-1,(x,y)\updownarrow} \end{bmatrix}, \quad (4.6)$$

where

$$U := P_+ + P_- + P_0 = \begin{bmatrix} \epsilon_+ & J & \chi \\ J & \epsilon_- & \chi \\ \chi & \chi & \epsilon_0 \end{bmatrix}, \quad \text{and} \quad [0] := \begin{bmatrix} 0 & 0 & 0 \\ 0 & 0 & 0 \\ 0 & 0 & 0 \end{bmatrix}. \quad (7, 8)$$

One can easily recognize from (4.6) that the two DPs (y_{DP+} and y_{DP-}) follow zigzag-shaped routes (represented by bent red and blue arrows). The matrix U of (7) meets a unitarity requirement for the QW model by setting to

$$U = \begin{bmatrix} \epsilon_+ & J & \chi \\ J & \epsilon_- & \chi \\ \chi & \chi & \epsilon_0 \end{bmatrix} = \begin{bmatrix} -\cos^2\theta & \sin^2\theta & \sin(2\theta)/\sqrt{2} \\ \sin^2\theta & -\cos^2\theta & \sin(2\theta)/\sqrt{2} \\ \sin(2\theta)/\sqrt{2} & \sin(2\theta)/\sqrt{2} & \cos(2\theta) \end{bmatrix}. \quad (4.9)$$

For dealing with the DP at the left border of the lattice, as an example, a matrix

$$\sigma = \begin{bmatrix} 0 & 1 & 0 \\ 1 & 0 & 0 \\ 0 & 0 & 1 \end{bmatrix}, \quad (4.10)$$

is introduced to represent the reflection of $y_{DP-\leftarrow}$ that hops along the $-x$ axis and is normally incident at this border (Fig. 4.2).

4.3 Dressed-Photon-Phonon Creation Probability on the Tip of a Fiber Probe

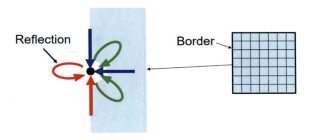

Fig. 4.2 Reflection at the left border

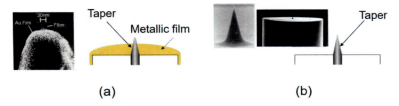

Fig. 4.3 External forms and cross-sectional structures of the fiber probe. **a** A high-efficiency fiber probe. **b** A basic fiber probe

4.3 Dressed-Photon-Phonon Creation Probability on the Tip of a Fiber Probe

Figure 4.3 shows the external forms and cross-sectional structures of fiber probes [4]. The DPP energy is lost at the taper of the fiber probe. This corresponds to radiating scattered light from the taper. In order to avoid this radiation, an opaque metallic film is formed on the taper (Fig. 4.3a) to realize a high-efficiency fiber probe. This is the prototype of devices that are now popularly used. Figure 4.3b is a basic fiber probe without a metallic film coating, resulting in DPP energy loss at the taper. This is a primitive device that has been used only in the early stages of DP science.

Figure 4.4a shows a right-angled isosceles triangular (RIT) lattice that approximates the profile of the fiber probe. It schematically explains that, by applying input signals to the sites on the base of the RIT lattice, DPs are created and transferred to the adjacent sites. During this transfer, DPPs are created by the DP-phonon coupling. These DPPs transfer through the RIT lattice and finally reach its apex (the tip of the fiber probe). This apex is assumed to be a sink from which the DPP energy is dissipated. This section calculates the creation probability P of the DPP at this sink [5, 6].

To cover a broader range of mathematical discussions based on the QW model, a phase angle ξ is introduced to the real-valued matrix in (7). As a result, U is replaced by a complex-valued matrix $U(\xi) = \exp(\xi)U$. The value of χ/J may be fixed to 1 for simplicity, indicating that the DP-phonon coupling energy is equal to

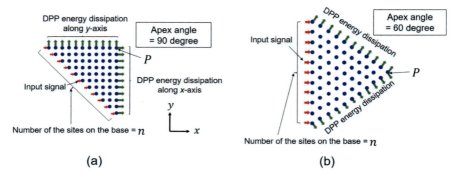

Fig. 4.4 Two-dimensional triangular lattices. **a** A right-angled isosceles triangle (RIT). **b** An equilateral triangle (ET)

Fig. 4.5 An example of the calculated temporal behavior of the value of P. T_p : Pulsation interval. T_s : The time required to converge to the stationary value

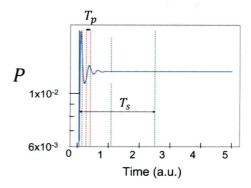

the DP hopping energy. However, to cover a broader range of physical discussions, the present section employs a wider range of values, i.e., $0.1 \leq \chi/J \leq 10$.

Figure 4.5 shows an example of the calculated temporal behavior of the value of P. After the input signals are applied to all the sites on the base of the RIT simultaneously, the value of P increases with time and reaches a stationary value.

4.3.1 Dependence on Parameters

For comparison with the experimental results derived by using the fiber probes of Fig. 4.3a, b, this subsection calculates the value of P without and with DPP energy loss at the taper, respectively, at the slope of the RIT lattice.

[Case 1: Without DPP energy loss]

Figure 4.6a shows the calculated dependence of P on ξ and χ/J (number of the sites on the base of the RIT lattice $n = 61$). Figure 4.6b shows the dependence of P on ξ at $\chi/J = 1$ in Fig. 4.6a. A lot of bumps are seen on the curve in this figure, which

4.3 Dressed-Photon-Phonon Creation Probability on the Tip of a Fiber Probe

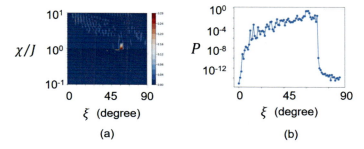

Fig. 4.6 The value of P in the case without DPP energy loss at the slope ($n = 61$). **a** Dependence on ξ and χ/J. **b** Dependence on ξ at $\chi/J = 1$ in (a)

are attributed to interference in the RIT lattice, originating from the reflection at the slope. This curve is symmetrical about $\xi = 90$ degree. Furthermore, the value of P in this figure takes the maximum P_{max} at $\xi = 67.5$ degree (=$(3/8)\pi$). It was found that P_{max} asymptotically approaches a constant value of 3×10^{-1} as n increases, from which it was confirmed that sufficiently high accuracy of approximation for numerical simulation was obtained when $n \geq 51$.

[Case 2: With DPP energy loss]

Figure 4.7a shows the calculated dependence of P on ξ and $\chi/J (= 61)$. Crescent-shaped red belts are seen in this figure, in which the value of P is very large in comparison with those outside the belts. The values of P show irregular variations and abrupt increases in the red belts and at their rim. Since these red belts are in the area $\chi/J > 1$, the value of P at $\chi/J = 1$ does not suffer any effects from the red belt.

Figure 4.7b shows the dependence of P on ξ at $\chi/J = 1$ in Fig. 4.7a. No bump on the curve is seen in this figure, which indicates that no interference takes place in the RIT lattice. This curve is symmetrical about $\xi = 90$ degree, as was the case

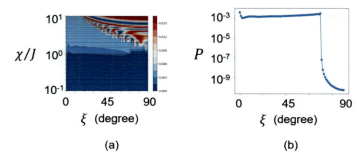

Fig. 4.7 The value of P in the case with DPP energy loss at the slope ($n = 61$). **a** Dependence on ξ and χ/J. **b** Dependence on ξ at $\chi/J = 1$ in (a)

of Fig. 4.6b. Furthermore, the value of P in this figure takes the maximum P_{max} at $\xi = 67.5$ degree ($=(3/8)\pi$). The value of P_{max} asymptotically approaches a constant value as n increases, from which it was confirmed that sufficiently high accuracy of approximation for numerical calculation was obtained when $n \geq 51$, as in the Case 1 above.

The value of P_{max} was 3×10^{-3} for $n \geq 51$, which is 10^{-2} times that in Fig. 4.6b. This indicates that it is effective to suppress the DPP energy loss (Case 1) at the slope of the RIT lattice to increase the probability of DPP creation. This indication is in agreement with experimental results in which the taper of a fiber probe is coated with an opaque metallic film to suppress the DPP energy loss and to increase the DPP creation efficiency at the tip of the fiber probe (Fig. 4.3a).

4.3.2 Dependence on the Apex Angle of a Fiber Probe

This subsection calculates the probability P at the apex of an equilateral triangular (ET) lattice. For reference, experimental studies have found that the value of P was smaller for smaller apex angles [7]. Unlike the 90 degree apex angle of the RIT lattice in Subection 4.3.1, this subsection deals with a triangular lattice with a smaller apex angle, i.e., an ET lattice (the apex angle of 60 degrees), as an example, as is shown by Fig. 4.4b. In the RIT lattice (Fig. 4.4a), each site has four nearest neighbor sites located along the $\pm x$- and $\pm y$- axes. This means that the DPP energy transfers from/to these four sites. In contrast, each site in the ET lattice (Fig. 4.4b) has six nearest-neighbor sites located along the directions of $\exp(\pm i\pi/6)x$-, $\exp(\pm i5\pi/6)x$-, and $\pm y$-axes. The DPP energy transfers from/to these six axes. By noting the number of these nearest-neighbor sites, the spatial-temporal evolution equation for the ET lattice was derived by modifying that for the RIT lattice, and the values of P was calculated for the case without and with DPP energy loss at the slope.

Figure 4.8a, b show the calculated dependence of P on ξ and χ/J without and with DPP energy loss at the slope, respectively at the slope of the ET lattice ($n = 51$).

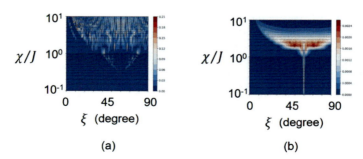

Fig. 4.8 Dependence of P on ξ and χ/J (ET lattice: $n = 51$). **a**, **b** are in the cases without and with DPP energy loss at the slope, respectively

4.4 Dressed-Photon-Phonon Confined by a B Atom-Pair in a Si Crystal

Table 4.1 Calculated values of P_{max}

	RIT	ET
Without DDP energy loss at the slope	3×10^{-1}	1×10^{-2}
With DDP energy loss at the slope	3×10^{-3}	1×10^{-3}

In Fig. 4.8b, the red belt is seen, as was the case in Fig. 4.7a. Since this belt is in the area $\chi/J > 1$, the value of P at $\chi/J = 1$ does not suffer any effects from this belt. Several results were derived from Fig. 4.8a, b, that are consistent with those of the RIT lattice. They are:

(a) The value of P_{max} asymptotically approaches a constant value as n increases, from which it was confirmed that sufficiently high accuracy of approximation for numerical simulation was obtained when $n \geq 51$.

(b) The value of P_{max} of Fig. 4.8a was 1×10^{-2} for $n \geq 51$, which is 10 times that of Fig. 4.8b (1×10^{-3}). This indicates that it is effective to suppress DPP energy loss at the slope of the ET lattice to increase the value of P, as was described in Subsection 4.3.1. This indication is in agreement with the experimental results.

Table 4.1 summarizes the calculated values of P_{max} for $\chi/J = 1$. This table shows that the value without DPP energy loss at the slope is larger than that with DPP energy loss, which is in agreement with experimental results. Furthermore, the value is larger for the RIT lattice (an apex angle of 90 degree) than that for the ET lattice (an apex angle of 60 degree), which is also in agreement with experimental results [7].

4.4 Dressed-Photon-Phonon Confined by a B Atom-Pair in a Si Crystal

By using a two-dimensional quantum walk (QW) model, this section analyzes the spatial distribution of a DPP that was confined by an impurity boron (B) atom-pair in a silicon (Si) crystal [3, 8–10]. It has been experimentally confirmed that, when the direction of the B atom-pair is perpendicular to that of the irradiated light propagation, the DPP creation probability is higher than the case of other orientation directions of the B atom-pair, including the parallel orientation [11, 12]. The vertical axis in Fig. 4.9 indicates the difference ΔN in the number of B atom-pairs (along the perpendicular direction) after and before the DPP-assisted annealing, as acquired by atom probe field ion spectroscopy with sub-nanometer spatial resolution. The horizontal axis indicates the ratio d/a between the length d of the pair and the lattice constant a of the Si crystal. Closed circles show that the value of ΔN takes the maximum at $d/a = 3$. It should be noted that the value of d/a at this maximum depends on the density of the B atoms doped prior to the DPP-assisted annealing, which decreases with increasing density of the doped B atoms.

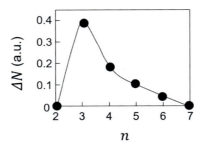

Fig. 4.9 Relation between the ratio d/a and the difference ΔN in the number of the B atom-pairs after and before the DPP-assisted annealing

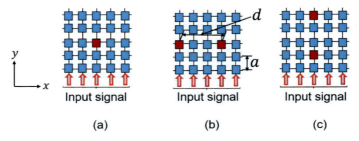

Fig. 4.10 Positions of B atoms in the square lattice of a Si crystal. Red and blue closed squares represent the sites of B and Si atoms, respectively. **a** A single B atom. **b, c** B atom-pairs oriented along the directions perpendicular and parallel to that of the incident light propagation (along the x- and y- axes), respectively

Numerical calculations are carried out below for the cases of a single B atom (Fig. 4.10a), a B atom-pair that is oriented along directions perpendicular (Fig. 4.10b), and parallel (Fig. 4.10c) to the direction of the incident light propagation. It should be noted that experiments have confirmed that the DPP creation probability is the highest in the case of Fig.4.10b.

The present section uses a QW model that is equivalent to that of Sect. 4.3. Since the DP couples with the phonon preferably at the B atom site, the value of χ at this site is fixed to be larger than J. On the other hand, $\chi = J$ at the Si atom site, as was the case in Sect. 4.3. The DPP reflection at the side of the square lattice can be neglected because the size of the Si crystal is much larger than the length of the B atom-pair. The present section does not require a complex-valued unitary matrix $U(\xi) = \exp(i\xi)U$ because the reflection of the DPP at the sides of the square lattice is neglected. To maintain the accuracy of approximating the Si crystal by the geometric square lattice model, the numbers n of the sites on the sides of the square lattice are increased to 51, as was employed in Sect. 4.3.

4.4 Dressed-Photon-Phonon Confined by a B Atom-Pair in a Si Crystal   51

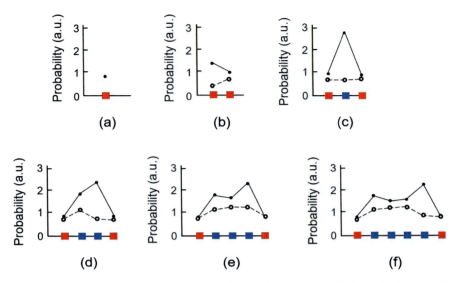

Fig. 4.11 Cross-sectional spatial distributions of the DPP creation probability in the B atom-pair along the *x*- and *y*-axes, represented by closed and open circles, respectively. Red and blue closed squares represent the sites of the B and Si atoms. The values of χ/J at the B atom is 20. The value $d/a = 0$ in **a** denotes that a single B atom exists in the lattice. **b–f** The value of d/a is 1 − 5

4.4.1 Dependence on the Direction of the B Atom-Pair

Figure 4.11 shows the calculated dependences of the stationary spatial distribution of the DPP creation probability $|\Psi_{t,(x,y)}|^2$ on the direction of the B atom-pair in the square lattice ($n = 51$). The value of χ/J at the B atom is fixed to be 20. Closed and open circles represent the spatial distributions of the DPP creation probability at the B atom-pairs oriented along the *x*- and *y*- axes, respectively. The value of d/a is varied from 1 to 5. For reference, the results for the cases of a single B atom ($d/a = 0$) are also shown. From the large values represented by closed circles, it is confirmed that the DPP is effectively confined by the B atom-pair along the *x*- axis. The small values represented by open circles indicate that the DPP is not effectively confined by the B atom-pair along the *y*- axis. These orientation-dependences agree with the experimental results reviewed above. Figure 4.11 also shows that the probability distributions, represented by the closed circles, are asymmetric. The origin of this will be discussed using Fig. 4.13.

Two measures (1) and (2) are used to evaluate the confinement above:
(1) The DPP creation probability C_{av}, averaged over all the sites in the B atom-pair: It is expressed as $C_{av} := A_{av} - B_{av}$. Here, A_{av} is the DPP creation probability, averaged over the Si atoms between the two B atoms. B_{av} is the average for the sites of the two B atoms.

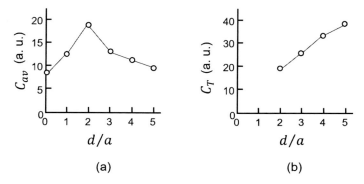

Fig. 4.12 The dependences of C_{av} and C_T on d/a. $\chi/J = 20$ at the B atom

(2) The DPP creation probability C_T, integrated over all the sites in the B atom-pair: It is expressed as $C_T = ((d/a) - 1)C_{av}$.

Figure 4.12a shows that the value of C_{av} takes the maximum at $d/a = 2$ and decreases monotonically with the increase of d/a, which agrees with the experimental results in Fig. 4.9.[1] Figure 4.12b shows that the value of C_T increases monotonically with the increase of d/a, which is due to the contribution from $(d/a) - 1$ in C_T.

4.4.2 Asymmetric Distribution and Photon Breeding

Since closed circles in Fig. 4.11 shows that DPP is effectively confined by the B atom-pair along the x-axis, the present subsection presents the calculated results only for this B atom-pair. Calculated probabilities $|y_{DP+}|^2$, $|y_{DP-}|^2$ and $|y_{Phonon}|^2$ at this B atom-pair are shown in Fig. 4.13a–c, respectively ($n = 51$, $d/a = 3$ and $\chi/J = 20$ at the B atoms). The value of $|y_{DP+}|^2$ (Fig. 4.13a) is larger than that of $|y_{DP-}|^2$ (Fig. 4.13b). This means that the DP creation probability $|y_{DP+}|^2$, travelling in the same direction as that of the incident light propagation (along the $+y$-axis), is larger. In other words, the DP creation probability, traveling in the same direction as that of the incident photon momentum, is larger. This indicates the presence of photon breeding (PB) with respect to the photon momentum [Phenomenon 13] [13].

Furthermore, the distributions of $|y_{DP+}|^2$ and $|y_{DP-}|^2$ are asymmetric in spite of the symmetric distribution of the phonon $|y_{Phonon}|^2$ (Fig. 4.13c). In Fig. 4.13 a, the value of $|y_{DP+}|^2$ at the Si atom on the right is larger than that on the left. In contrast, in Fig. 4.13b, the value of $|y_{DP-}|^2$ at the Si atom on the left is larger. From this

---

[1] Although Fig. 4.9 shows that the value of ΔN takes the maximum at $d/a = 3$, this maximum value is apt to fluctuate, depending on the density fluctuations of the B atoms doped into the Si crystal prior to the DPP-assisted annealing. Thus, it can be claimed that Fig. 4.12a agrees with Fig. 4.9.

4.4 Dressed-Photon-Phonon Confined by a B Atom-Pair in a Si Crystal

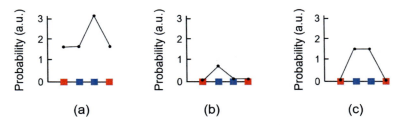

Fig. 4.13 Cross-sectional spatial distributions of the creation probabilities of DP and phonon. $d/a = 3$ and $\chi/J = 20$ at the B atom (red closed square). **a**, **b** are for $|y_{DP+}|^2$ and $|y_{DP-}|^2$, respectively. **c** is for $|y_{Phonon}|^2$

contrast, it is concluded that the asymmetry in $|\Psi_{t,(x,y)}|^2$ (closed circles in Fig. 4.11) originates from that in $|y_{DP+}|^2$ of Fig. 4.13a.

4.4.3 A Quantum Walk Model with Energy Dissipation

It should be pointed out that the experiments on the Si-light emitting devices (Sect. 2.4.1 and 2.4.2) measured not the energy of the DPP itself confined in the B atom-pair but the on-shell field energy that leaked out from the Si crystal. This energy was created due to the dissipation of the off-shell DPP field energy. For comparison with experimental results, this subsection introduces the phenomenological dissipation constant κ into the QW model [3, 9].

Novel equations are presented to deal with the energy dissipation by using a four-dimensional vector

$$\Psi'_{t,(x,y)} = \begin{bmatrix} y_{DP+} \\ y_{DP-} \\ y_{Phonon} \\ y_{dis} \end{bmatrix}_{t,(x,y)} \quad (4.11)$$

y_{DP+} and y_{DP-} are the creation probability amplitudes of the DPs that travel by repeating the hopping in the upper-right and lower-left directions, respectively, and y_{Phonon} is that of the phonon. y_{dis} represents the creation probability amplitude of the energy dissipated from the square lattice. The spatial-temporal evolution equation is represented by

$$\Psi'_{t,(x,y)} = \begin{bmatrix} \Psi'_{t,(x,y)\leftrightarrow} \\ \Psi'_{t,(x,y)\updownarrow} \\ \Psi''_{t,(x,y),dis} \end{bmatrix} = \begin{bmatrix} [0] & \hat{U} & [0] \\ \hat{U} & [0] & [\kappa] \\ [0] & [\kappa] & [0] \end{bmatrix} \begin{bmatrix} \Psi'_{t-1,(x,y)\leftrightarrow} \\ \Psi'_{t-1,(x,y)\updownarrow} \\ \Psi''_{t-1,(x,y),dis} \end{bmatrix}, \quad (4.12)$$

where $\hat{U} := \sqrt{1 - \kappa^2}U$. The third line of the vector corresponds to the dissipated energy

$$\Psi''_{t,(x,y),dis} := \begin{bmatrix} y_{DP+,dis} \\ y_{DP-,dis} \\ y_{Phonon} \end{bmatrix}_{t,(x,y)} . \tag{4.13}$$

The first and second lines ($y_{DP+,dis}$ and $y_{DP-,dis}$) in (4.13) represent the dissipated energies that travel along the directions parallel and anti-parallel to that of the irradiated light, respectively. Their sources are y_{DP+} and y_{DP-} in (4.11). The phonon (y_{Phonon}) does not contribute to the energy dissipation due to its non-travelling nature.

Since the DP couples with the phonon preferably at the B atom site, χ/J has to be larger than unity. Here, it is fixed to 20[7]. On the other hand, $\chi = J$ at the Si atom sites.

The quantity κ in the matrix

$$[\kappa] := \begin{bmatrix} \kappa\ 0\ 0 \\ 0\ \kappa\ 0 \\ 0\ 0\ 0 \end{bmatrix} \tag{4.14}$$

on the right-hand side of (4.12) is a phenomenological dissipation constant ($0 \leq \kappa \leq 1$). The quantity $\sqrt{1 - \kappa^2}$ represents the magnitude of the energy left in the square lattice after energy dissipation.

By referring to Sect. 4.3.1, numerical simulation was carried out for the case of the B atom-pair oriented along the x-axis. Its length d was fixed to $3a$. The creation probability A_{av}, averaged over the Si atoms in the B atom-pair, was used for evaluating the magnitude of the confinement. Although Sect. 4.2 claimed that the number n of the sites on the side of the square lattice must be equal to or higher than 51, n is fixed to 21 in the present subsection for shortening the calculation time of the computer system. By preliminary calculations, it was confirmed that the calculated results for $n = 21$ and $n = 51$ are equivalent to each other. The origins of this equivalence are: In the case of a fiber probe, a complicated triangular lattice was required, and the reflection of the DPP had to be taken into account at the slope of the triangle (Sect. 4.2). In contrast, a simpler square lattice was accepted in the present study. In addition, it was permissible to neglect the reflection at the sides of the square lattice [10].

Figure 4.14 shows the relation between κ and A_{av} of the energy that dissipated from the B atom-pair and leaked out to the external space. It is the magnitude of the dissipated energy $|y_{DP+,dis}|^2$ in (4.13) that travels along the direction parallel to that of the irradiated light propagation along the $+y$-axis. This figure shows that A_{av} takes the maximum at $\kappa = 0.2(:= \kappa_{opt})$. The existence of such an optimum value κ_{opt} agrees with experimental results and implies the feature of "comfortability"that has been studied using the QW model [14].

Since the dissipated energy is the on-shell field energy that emerges from the off-shell field, it spreads over the whole volume of the Si crystal and finally leaks out from

4.4 Dressed-Photon-Phonon Confined by a B Atom-Pair in a Si Crystal

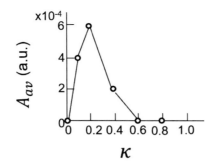

Fig. 4.14 Relation between κ and A_{av} of the dissipated energy $|y_{DP+,dis}|^2$

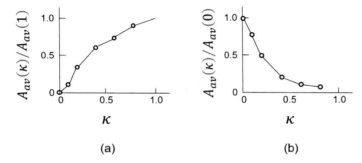

Fig. 4.15 Relation between κ and the total energy. **a**, **b** are for the total dissipated energy and the total energy of the source of dissipation, respectively

the Si crystal. Thus, the magnitude of the total energy was calculated by summing up over all the sites of the square lattice. Figure 4.15a shows such total dissipated energy. This figure shows that A_{av} monotonically increases with κ, which implies that the total dissipated energy increases with the increase of κ and spreads over the whole volume of the Si crystal. This spread corresponds to diffraction, which is an intrinsic feature of the on-shell field in a visible macroscopic space. Figure 4.15b represents the values of A_{av} for the total energy of the source of dissipation. It shows that A_{av} monotonically decreases with κ. By comparing Fig. 4.15a, b, the complimentary feature was confirmed to meet the energy conservation requirement.

In contrast to the complimentary feature in the visible macroscopic space above, the existence of the optimum value κ_{opt} in Fig. 4.14 implies the intrinsic feature of the off-shell DPP field in an invisible microscopic space. By this contrast, it is confirmed that the present QW model with energy dissipation could successfully describe the intrinsic features of invisible microscopic and visible macroscopic spaces in a consistent manner.

Figure 4.13 discussed the photon breeding (PB) with respect to the photon momentum by using the asymmetric spatial distribution of the DPP creation probability as

Fig. 4.16 Creation probability of the dissipated energy ($\kappa = 0.2$) at the sites in the B atom-pair. **a**, **b** are the dissipated energies $|y_{DP+,dis}|^2$ and $|y_{DP-,dis}|^2$, respectively. Red and blue squares represent the sites of the B and Si atoms, respectively

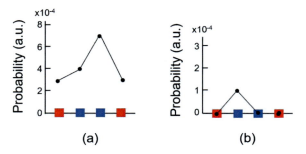

a clue. On the other hand, since experiments were performed to evaluate the characteristics of the light that leaked out from the Si crystal for discussing PB, the QW model with energy dissipation is used here for a more detailed comparison with experimental results.

Figure 4.16a shows the probability for the dissipated energy $|y_{DP+,dis}|^2$ ($\kappa = 0.2$). This figure shows that the probability at the right site of the Si atom is larger than that at the left site. Figure 4.16b is for the dissipated energy $|y_{DP-,dis}|^2$ ($\kappa = 0.2$) whose probability at the left site of the Si atom is larger than that at the right site. Here, it should be noted that the probability in Fig. 4.16a is much larger than that in Fig. 4.16b. That is, the magnitude of the dissipated energy $|y_{DP+,dis}|^2$ (traveling in the direction parallel to that of the incident light) is much larger than that of $|y_{DP-,dis}|^2$ (traveling in the anti-parallel direction). It implies that the momentum of the photon, created due to the dissipation and leaked out from the Si crystal, is equivalent to that of the incident light. This equivalence manifests the PB with respect to the photon momentum.

Figure 4.17a, b show the relations between κ and the creation probabilities of the dissipated energies $|y_{DP+,dis}|^2$ and $|y_{DP-,dis}|^2$ in (4.13), which travel in the directions parallel and anti-parallel to that of the incident light, respectively. The values of A_{av} for $|y_{DP+,dis}|^2$ (Fig. 4.17a) are larger than those for $|y_{DP-,dis}|^2$ (Fig. 4.17b), which represents the PB with respect to the photon momentum, as was indicated in Fig. 4.16. For reference, the values of A_{av} in Fig. 4.17a, b take the maxima at $\kappa = 0.2$, which is in agreement with Fig. 4.14. From a discussion of Figs. 4.16 and 4.17, it

Fig. 4.17 The relations between κ and the creation probability of the dissipated energy. **a**, **b** are for $|y_{DP+,dis}|^2$ and $|y_{DP-,dis}|^2$, respectively

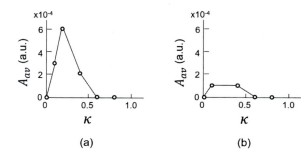

4.5 Photon Breeding with Respect to Photon Spin   57

is concluded that the PB with respect to the photon momentum was successfully confirmed by the QW model with energy dissipation.

The present subsection introduced an energy dissipation constant κ into the two-dimensional QW model for detailed analysis of the DPP creation and confinement by a B atom-pair in a Si crystal. As a result, it succeeded in describing the intrinsic features of the DPP energy transfer, including:

(a) The magnitude of the energy dissipated from the B atom-pair took a maximum at $\kappa = 0.2$.

(b) The total dissipated energy over the whole volume of the Si crystal monotonically increased with κ. On the other hand, the total energy of the source of dissipation complementarily decreased.

(c) The dissipated energy exhibited the feature of photon breeding (PB) with respect to the photon momentum.

Among the features above, (a) implies localization of the DPP, which is an intrinsic feature of an off-shell field in a microscopic space. In contrast, (b) represents the feature of the dissipated energy of the macroscopic on-shell field. It was confirmed from these contrasting features that the intrinsic features of microscopic and macroscopic fields were successfully described by introducing the energy dissipation into the QW model in a consistent manner.

At the end of this section, it should be pointed out that the constant κ is a phenomenological quantity that was introduced to analyze the energy dissipation. It is expected that its identity can be revealed by future studies of classical and quantum measurement theories [15].

4.5 Photon Breeding with Respect to Photon Spin

Relevant experimental results on Si-LEDs, to be analyzed here, are summarized as follows [11, 12]: Although the method of fabrication was equivalent to the one reviewed in Sect. 2.4.1, linearly polarized light was used for the DPP-assisted annealing. After fabrication, a current was injected to the device for LED operation. The degree of polarization P [16] of the emitted light was found to be nonzero at the photon energy $h\nu_{anneal}$ of the light irradiated for the DPP-assisted annealing. The value of P increased by increasing the DPP-assisted annealing time and saturated to $7 \times 10^{-2} (= 7\%)$. These experimental results confirmed that the light emitted from the fabricated Si-LED was polarized, and the polarization direction was governed by that of the light irradiated during the DPP-assisted annealing, that is to say, PB with respect to the photon spin.

The phenomenon to be analyzed in the present section is schematically summarized in Fig. 4.18, by referring to the experimental results above [17]. Here, the light propagates along the z-axis and is incident on the Si crystal, as was the case of the light irradiated during the DPP-assisted annealing. It is assumed that the B atom-pair orients along the x-axis because from experiments it was found that the DPP creation probability at this pair is higher than the case when it is oriented along the z-axis (Sect. 4.4.1).

Furthermore, as shown in Fig. 4.18a, the incident light is assumed to be linearly polarized along the y-axis. Experiments have confirmed that this polarized light realized higher creation/confinement probability of the DPP at the B atom-pair compared with the case of light linearly polarized along the x-axis (Fig. 4.18b) [11, 12]. The goal of the present section is to reproduce this experimental confirmation by numerical calculations using a three-dimensional QW model [17].

4.5.1 Three-Dimensional Quantum Walk Model

Figure 4.19 represents a three-dimensional lattice that is used as the three-dimensional QW model of the Si crystal. Blue squares represent the sites of Si atoms. Two red squares represent a B atom-pair at the center of the lattice. By referring to the calculated result in Sect. 4.4.1, it is assumed that this pair orients along the x-axis, and its length d is three times the lattice constant a of the Si crystal ($d = 3a$).

In the case of the three-dimensional lattice, the bent arrows in Fig. 4.1b, c are replaced by red and blue bent arrows in Fig. 4.20. The senses of the repetitive DP hopping in Fig. 4.20a, b are those of advancing right-handed and left-handed screws, respectively. They correspond to modes 12 and 1, respectively, in [2].

DP is created at each site on the first bottom layer (Fig. 4.19) of the three-dimensional lattice by the incident light that propagates along the $+z$-axis. The created DP subsequently hops to the nearest-neighbor site. In order to introduce the polarization of the incident light in the QW model, first, the case of linear polarization along the y-axis (Fig. 4.18a) is dealt with. Since the electric field of this light oscillates along the y-axis, classical electromagnetics teaches that this electric field creates an electric dipole in the NP or atom that oscillates also along the y-axis (Fig. 4.21a). This oscillating dipole emits an electromagnetic field that propagates along

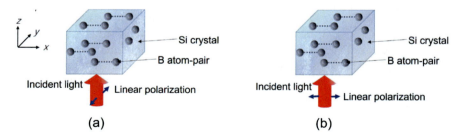

Fig. 4.18 Relations among the directions of the orientation of a B atom b-pair (along the x-axis), the incident light propagation (the z-axis), and its polarization. **a**, **b** are for the linear polarization along the y- and x- axes, respectively

4.5 Photon Breeding with Respect to Photon Spin

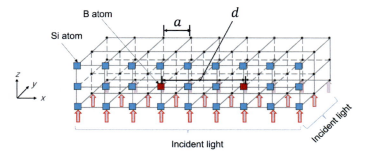

Fig. 4.19 Blue and red squares represent the sites of the Si and B atoms, respectively. The length d of the B atom-pair, oriented along the x-axis, is $d = 3a$, where a is the lattice constant of the Si crystal

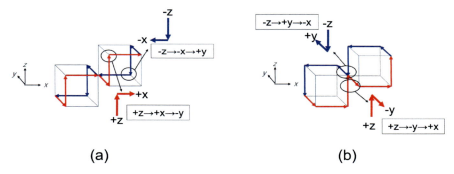

Fig. 4.20 DP hopping from one site to its nearest-neighbor in the three-dimensional lattice. **a**, **b** The senses of the repetitive DP hopping are those of advancing right-handed and left-handed screws, respetively

the x-axis. Thus, it can be assumed that the DP on the first bottom layer of the three-dimensional lattice hops along the x-axis (the first step in Fig. 4.22a). Here, since the incident light is an alternating electromagnetic field, it should be noted that the DP hops not only in the $+x$-axis direction but also in the $-x$-axis direction, alternately in time. By following red arrows in Fig. 4.20a, the DP subsequently hops along the y- and z-axes (the second and third steps, respectively, in Fig. 4.22a). In the third step, by hopping along the z-axis, the DP transfers energy from the sites in the first bottom layer to those in the second bottom layer. By repeating this hopping, DP energy transfers from the lower to the upper layers of the three-dimensional lattice and reaches the B atom-pair. Since hopping along the $-y$-axis and $-z$-axis is also allowed, as was the case of the $-x$ -axis above, the transfers from the upper to the lower layers are represented by blue arrows in Fig. 4.20a.

Second, in the case of the incident light with linear polarization along the x-axis (Fig. 4.18b), the created electric dipole oscillates also along the x-axis, and the emitted electromagnetic field propagates along the y-axis (Fig. 4.21b). Thus, the DP

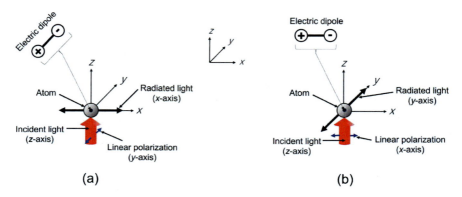

Fig. 4.21 Relations among the directions of the incident light propagation (the z-axis), its polarization, and the light propagation emitted from the electric dipole in a nanometer-sized particle (NP) or an atom. **a**, **b** are for the incident light that is linearly polarized along the y- and x-axes, respectively

starts hopping along the y-axis (the first step in Fig. 4.22b). Figure 4.22b also shows subsequent hopping of the second and third steps.

Numerical calculations were carried out to derive the stationary value of the creation probability of the DPP at the B atom-pair after repeating the DP hopping in Fig. 4.22. Since the B atom-pair in Fig. 4.18 orients along the x-axis, it is expected that the calculated value of the DPP creation probability in Fig. 4.22a is larger than that in Fig. 4.22b, as was experimentally confirmed. (4.2) and (4.4) can be also used for the spatial-temporal evolution equations for the state vector $\Psi_{t,(x,y)}$. These equations are governed by a unitary matrix that is represented by (7). The ratio χ/J is assumed to be 20 at the B atom site. The number of sites along a side of the three-dimensional lattice is set to 21, and thus, the total number of sites is $\{(21)^3\}$.[2] The reflection coefficient σ in (4.10) is set to 0 because the side length of the lattice is sufficiently longer than that of the B atom-pair.

4.5.2 Degree of Photon Breeding

Figure 4.23a shows calculated stationary values of $|\Psi_{t,(x,y)}|^2$ at the B atom-pair. Closed and open circles are the values for the cases of Fig. 4.22a, b, respectively. Figure 4.23b schematically defines the area in Fig. 4.23a that is surrounded by the horizontal axis and these circles connected by segments. It is expressed as

---

[2] It is known that the number density of the Si atoms in a Si crystal is about 10^{23} (m^{-3}). On the other hand, experiments confirmed that the doping density of B atoms in the Si crystal was about 10^{19} (m^{-3}). The ratio of these is about 10^4. The total number of sites in the lattice $(21)^3$ is nearly equal to this ratio. A super-computer was used to reduce the processing time of the massive amount of calculation data.

4.5 Photon Breeding with Respect to Photon Spin

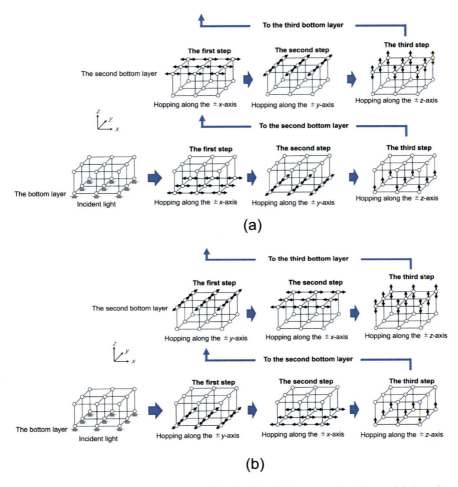

Fig. 4.22 Relations among the directions of the incident light propagation (the z-axis), its polarization, and the light propagation emitted from the electric dipole in a nanometer-sized particle (NP) or an atom. **a**, **b** are for the incident light that is linearly polarized along the y- and x-axes, respectively

$$A = \frac{a}{2}[b_1 + 2(s_1 + s_2) + b_2]. \tag{4.15}$$

By referring to the formula for the degree of polarization P of [16], the degree of PB is defined as

$$DoPB = \frac{A_{\rightarrow (z \rightarrow x)} - A_{\uparrow (z \rightarrow y)}}{A_{\rightarrow (z \rightarrow x)} + A_{\uparrow (z \rightarrow y)}}, \tag{4.16}$$

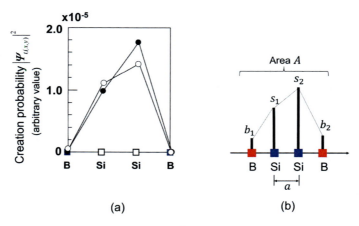

Fig. 4.23 Calculated stationary values $|\Psi_{t,(x,y)}|^2$ of DPP at the B atom-pair. **a** Closed and open circles represent the calculated value for Fig. 4.22a, b, respectively. **b** Schematic definition of the area A

Table 4.2 The calculated values of $DoPB$

$|\Psi_{t,(x,y)}|^2$	$DoPB_\Psi = 2.0 \times 10^{-2}\ (= 2.0\%)$
$|y_{DP+}|^2$	$DoPB_{DP+} = 1.9 \times 10^{-1}\ (= 19\%)$
$|y_{DP-}|^2$	$DoPB_{DP-} = -4.3 \times 10^{-1}\ (= -43\%)$
$|y_{Phonon}|^2$	$DoPB_{Phonon} = 7.3 \times 10^{-2}\ (= 7.3\%)$

where $A_{\rightarrow(z\rightarrow x)}$ and $A_{\uparrow(z\rightarrow y)}$ are the areas surrounded by the closed and open circles in Fig. 4.23a, respectively. By using Fig. 4.23a, (4.16) derives the value of $DoPB_\Psi$ for $|\Psi_{t,(x,y)}|^2$ as $2.0 \times 10^{-2} (= 2\%)$. This positive value indicates that the DPP creation probability in Fig. 4.18a (and Fig. 4.22a) is higher than that in Fig. 4.18b (and Fig. 4.22b). The value $2.0 \times 10^{-2} (= 2\%)$ falls in the range of the experimentally evaluated values of $P (0 \leq P \leq 7 \times 10^{-2})$. These features indicate PB with respect to the photon spin and confirm that the calculated results above agree with the experimental results. Thus, it is concluded that the goal of the present section was achieved.

Figure 4.24a–c show calculated stationary values of $|y_{DP+}|^2$, $|y_{DP-}|^2$ and $|y_{Phonon}|^2$ at the B atom-pair. The values of $DoPB_{DP+}$, $DoPB_{DP-}$ and $DoPB_{Phonon}$ are derived from these figures and (4.16) and are presented in Table 4.2. It should be pointed out that y_{DP+} represents the probability amplitude of the DP that hops in the direction parallel to that of the incident light propagation. Furthermore, the positive $DoPB_{DP+}$ in this table indicates that the creation probability of the component y_{DP+} is higher in the case of Fig. 4.18a than of Fig. 4.18b. On the other hand, y_{DP-} represents the probability amplitude of the DP that hops in the direction anti-parallel to the that of the incident light propagation. The negative $DoPB_{DP-}$ indicates that the creation probability of the component y_{DP-} is higher in the case of Fig. 4.18b. Even though $DoPB_{DP-}$ is about 2.3 times larger than that of $DoPB_{DP+}$, $DoPB_\Psi$

4.5 Photon Breeding with Respect to Photon Spin

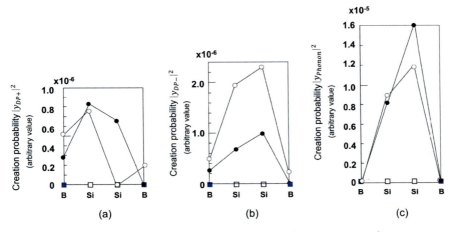

Fig. 4.24 Calculated stationary values $\left((a)|y_{DP+}|^2, (b)|y_{DP-}|^2, \text{ and }(c)|y_{Phonon}|^2\right)$ at the B atom-pair for the constituent elements (y_{DP+}, y_{DP-}, and y_{Phonon}) of $\Psi_{t,(x,y)}$

takes a positive value (2.0×10^{-2} ($= 2\%$)). This again indicates PB with respect to the photon spin and, furthermore, it confirms that the electromagnetic nature of $\Psi_{t,(x,y)}$ mainly depends on that of y_{DP+}.

Since the light originated from y_{DP+} propagates in a direction parallel to that of the incident light, and since this parallel propagation feature corresponds to PB with respect to the photon momentum, the indication and confirmation noted above imply that PB with respect to the photon spin and PB with respect to momentum are correlated. Furthermore, this correlation suggests that, if the energy dissipation constant of (4.14) could be introduced into the present three-dimensional QW model in the future, the value of $DoPB_\Psi$ could be evaluated by using the polarization feature of the observable macroscopic light that is emitted from the Si crystal. It is expected that this evaluation could derive a more accurate value of $DoPB_\Psi$ that could be larger than that in Table 4.2.

The present section assumed that the direction of the energy transfer of the DP was identical to the propagation direction of the light emitted from the created electric dipole (Figs. 4.21 and 4.22). However, it should be examined more carefully whether this assumption is fully acceptable from the viewpoint of off-shell science. Future studies based on novel theories of light-matter interactions in a microscopic space (refer to Chaps. 5–10) could find the answers to the fundamental question: "Are these teachings from classical electromagnetics acceptable for describing the phenomena in a nanometer-sized space?"

References

1. Konno, N.: Quantum walk. In: Franz, U., Schürmann, M. (Eds.), Quantum Potential Theory, pp. 309–452. Springer, Heidelberg (2008)
2. Ohtsu, M.: A Quantum Walk Model for Describing the Energy Transfer of a Dressed Photon. Off-shell Archive (2021) OffShell: 2109R.001.v1. https://rodrep.or.jp/en/off-shell/review_2109R.001.v1.html
3. Ohtsu, M., Segawa, E., Yuki, K., Saito, S.: A quantum walk model with energy dissipation for a dressed-photon-phonon confined by an impurity atom-pair in a crystal. OffShell: 2304O.001.v1. https://rodrep.or.jp/en/off-shell/original_2304O.001.v1.html
4. Ohtsu, M.: Off-Shell Applications In Nanophotonics, pp. 8–9. Elsevier, Amsterdam (2021)
5. Ohtsu, M., Segawa, E., Yuki, K.: A Quantum Walk Model for the Energy Transfer of a Dressed Photon. Abstracts of the 2022 International Symposium on Nonlinear Theory and its Applications (NOLTA2022), December 12–15, (Online meeting), paper number A3L-C1, pp. 65–68 (2022)
6. Ohtsu, M., Segawa, E., Yuki, K., Saito, S.: Dressed-photon-phonon creation probability on the tip of a fiber probe calculated by a quantum walk model. Off-shell Archive (December, 2022) OffShell: 2212O.001.v1. https://rodrep.or.jp/en/off-shell/original_2212O.001.v1.html
7. Ohtsu, M.: Near-Field Nano/Atom Optics and Technology, pp. 71–87. Springer, Tokyo/Berlin (1998)
8. Ohtsu, M., Segawa, E., Yuki, K., Saito, S.: Quantum walk analysis of spatial distribution of dressed-photon-phonon. 10th International Congress on Industrial and Applied Mathematics (August, 2023, Tokyo)
9. Ohtsu, M., Segawa, E., Yuki, K., Saito, S.: A quantum walk model with energy dissipation for dressed-photon-phonon confined by an impurity atom-pair in a crystal. 2023 International Symposium on Nonlinear Theory and Its Applications, (September, 2023, Italy)
10. Ohtsu, M., Segawa, E., Yuki, K., Saito, S.: Spatial distribution of dressed-photon-phonon confined by an impurity atom-pair in a crystal. Off-shell Archive, OffShell: 2301O.001.v1 (January, 2023). https://rodrep.or.jp/en/off-shell/original_2301O.001.v1.html
11. Kawazoe, T., Nishioka, K., Ohtsu, M.: Polarization control of an infrared silicon light-emitting diode by dressed photons and analyses of the spatial distribution of doped boron atoms. Appl. Phys. A 121(4), 1409–1415 (2015)
12. Ohtsu, M.: Silicon Light-Emitting Diodes and Lasers, pp. 35–42. Springer, Heidelberg (2016)
13. Ohtsu, M.: Silicon Light-Emitting Diodes and Lasers, pp. 1–13. Springer, Heidelberg (2016)
14. Higuchi, K., Komatsu, T., Konno, N., Morioka, H., Segawa, H.: A Discontinuity of the Energy of quantum walk in impurities. Symmetry 13, 1134 (2021), https://doi.org/10.3390/sym13071134
15. Okamura, K.: Towards a measurement theory for off-shell quantum fields. Symmetry 13, 1183 (2021). https://doi.org/10.3390/sym13071183
16. Born, M., Wolf, E.: Principles of Optics, Fourth edition, p. 45. Pergamon Press, Oxford (1970)
17. Ohtsu, M., Segawa, E., Yuki, K., Saito, S.: Analyses of photon breeding with respect to photon spin by using a three-dimensional quantum walk model. Off-shell Archive (November, 2023) OffShell: 2311O.001.v1. https://rodrep.or.jp/en/off-shell/original_2311O.001.v1.html

Chapter 5
Introductory Remarks on Theoretical Chapters 6–10

Abstract Based on the preceding experimental studies on enigmatic nanoscale light field, a tentative basic concept of dressed photon (DP) was formulated around 1990s, and since then, the study on DP has been accelerated to accomplish fruitful optical engineering outcomes so far explained in the previous Chaps. 1–4. Needless to say, the initial effort of forming a theoretical model of DP was based on a trial-and-error approach. It was not until 2017 that a quite novel *off-shell approach in quatum field theory (QFT)* was employed to construct a heuristic model of DP. Chaps. 5–10, which constitute the latter half of this book, are reserved to explain the cutting edge of those novel theoretical studies on DP. In order to understand new studies, some pieces of interdisciplinary knowledge are required, so that we begin this introductory chapter with a couple of simple comments on off-shell quantum field and on the other prerequisite knowledge on field dynamics.

5.1 On Off-Shell Quantum Fields

A major goal of our research on dressed photon studies is to reexamine the basic concepts of (relativistic) quantum field theory (QFT) of which present status is far from completion. We can safely say that the prevalent concept on quatum field is highly restricted one in the sense that it is based largely on a non-interacting free field which can be interpreted as the collection of on-shell particles $\{p^\mu(i); \ i = 1, 2, \ldots\}$, each of which satisfying the on-shell condition on momentum $p(i)^\mu$:

$$p(i)^\nu p(i)_\nu = [m(i)c]^2 \geq 0, \tag{5.1}$$

where $p^\mu(i)$ and $m(i)$ respectively denote the momentum and mass of i-th particle with c being light velocity and ν represents coordinate index of four dimensional (4D) space–time. In Eq. (5.1), we employ Einstein summation convention for ν and the sign convention of $(+, -, -, -)$ for Lorentzian metric tensor.

As we readily find in standard texts on QFT, the essential aspect of the prevalent QFT is that a given quantum field $\hat{\varphi}(x)$ can be described in terms of a multi-particles system which is a bundle of free particles satisfying the on-shell condition (5.1).

© The Author(s), under exclusive license to Springer Nature Switzerland AG 2025 65
M. Ohtsu and H. Sakuma, *Dressed Photons to Revolutionize Modern Physics*,
Nano-Optics and Nanophotonics, https://doi.org/10.1007/978-3-031-77944-2_5

Actually, by the mode decomposition of Fourie transformation, we have

$$\hat{\varphi}(x) = \int \frac{d^3 p}{(2\pi)^3 \sqrt{\vec{p}^2 + m^2 c^2}} \left[a(\vec{p}) \exp{(-i x^0)} \sqrt{\vec{p}^2 + m^2 c^2} + i \vec{p}\vec{x}) \right]$$

$$+ \int \frac{d^3 p}{(2\pi)^3 \sqrt{\vec{p}^2 + m^2 c^2}} \left[a^*(\vec{p}) \exp{(i x^0)} \sqrt{\vec{p}^2 + m^2 c^2} - i \vec{p}\vec{x}) \right]$$

$$= \int \frac{d^4 p}{(2\pi)^3} \delta(p^2 - m^2 c^2) [a(\vec{p}) \exp{(-i p x)} + a^*(\vec{p}) \exp{(i p x)}], \quad (5.2)$$

where we have $[a(\vec{p}), a^*(\vec{q})] = \delta(\vec{p} - \vec{q})$ and $\hat{\varphi}(x)$ satisfies Klein Gordon (KG) equation of the form

$$(\partial^\nu \partial_\nu + m^2 c^2)\hat{\varphi}(x) = 0. \quad (5.3)$$

Furthermore, if we add the concept of vacuum state $|0\rangle$ having 0 particle to the above system (5.2), then the elaborate scheme called *second quantization* makes it possible to represent multi-particle states $|\vec{p}_1, \ldots, \vec{p}_n\rangle$ together with the increase and decrease of the involved particles through a single quantum field $\hat{\varphi}(x)$.

The impact of second quantization on many of the physicists seemed to be so great for them to erroneously believe that the outcome of it describes fundamental characteristics of general qunatum fields. Notice, however, that Eq. (5.3) is what we call a linear (non-interacting) free mode equation, so that the on-shell condition given by Eq. (5.1) relating also to the second quantization must be regarded as a property of restricted quantum field represented by Eq. (5.3). Another no less influential factor elicitting the misunderstanding is that superluminous spacelike momentum field p^μ violating the on-shell condition (5.1) is usually regarded as a non-physical "ghost mode" in mainstream particle physics mainly because a "particle" associated with such a field breaks Einstein causality and partly because particle physicists' way of theoretical evaluation on their particle-collision experiments heavily depends upon Lehmann-Symanzik-Zimmermann (LSZ) reduction formulae on S-matrix which focuses only on the on-shell aspects of quantum fields in terms of time-ordered correlation functions by circumventing off-shell Heisenberg fields at the center of given quantum field interactions.

Although the notion of wave-particle duality revealed by quantum physics was certainly a groundbreaking finding, it does not necessarily mean that any physical entity can always be represented as such a dual-form. As a simple example, we can point out a well-known electromagnetic Coulomb mode which corresponds to a longitudinal electromagnetic field to be removed in conventional quatum electrodynamics (QED). The reason why this mode is eliminated in QED is simply because we cannot treat it as a particle field like usual transverse photon. In the extended discussion on Nakanish-Lautrup formalism [1] of abelian gauge theory, Ojima [2] clearly pointed out that Coulomb mode eliminated from "a particle form" in QED survives physically "as a non-particle wave form" to play an important role in electromagnetic field interactions.

5.1 On Off-Shell Quantum Fields

In order to describe physical events, we usually employ a space-time as "the platform stage" for the description of targeted objects. When those objects are represented by a finite set of particles, then the space-time we employ is usually called a coordinate system which does not have any important physical meaning. In the case of describing continuous objects or media, we have to introduce a notion of *field* for a given physical system, though the role played by a coordinate system remains to be a mathematical one. The situation changes drastically in the study of the theory of general relativity (TGR) where the curvature of space-time as a mathematical concept is tightly connected with the energy-momentum density of a given physical sytem. Notice, however, that since TGR belongs to the classical physics, 4 momentum p^μ in TGR satisfies the on-shell condition (5.1).

Literally speaking, the existence of off-shell momentum field:

$$p(i)^\nu p(i)_\nu = [m(i)c]^2 < 0, \tag{5.4}$$

means superluminous momentum field through which an instantaneous connection like quatum entanglement is realized. Historically, the problem on breaking Einstein causality by quantum entanglement state was first pointed out by Einstein, Podolsky and Rosen (EPR) as EPR paradox [3] and was finally settled by a series of experimental studies by Aspect's group. In relation to our main research theme of off-shell quantum fields, it is important to recognize the different characteristics existing between *physical quantities* under consideration and *states* of a given system which, according to *Micro-Macro Duality* (MMD) theory of Ojima [4], can be interpreted as the interface of micro-quantum and macro-classical systems. In case of fermionic fields, for example, we can observe the *state* change of a given fermionic field, while we cannot directly measure the fermionic *physical quantity* due to its anti-commutation relation leading to the violation of Einstein causality.

As we will see in Chap. 7 on spacelike electromagnetic field leading to dark energy field, an off-shell spacelike momentum field p^μ satisfying (5.4) actually arises from a certain fermionic field. In that case, the statement predicating that *such an off-shell momentum field p^μ is non-physical* is misleading since it simply means that it is an *unobservable physical quantity*. Notice that the notion of unobservable physical quantity is already employed in particle physics as the well-known notion of virtual particles playing the important roles in quatum field interactions. So that the aim of our new proposal on off-shell quantum fields is not to criticize the use of certain terminologies employed in standard quantum field theories but to try to improve our understanding on quantum field interactions from the viewpoint of Greenberg-Robinson (GR) theorem [5] derived in the field of axiomatic quantum field theory, which states clearly that *if the Fourier transform ($\varphi(p)$) of a given quantum field ($\varphi(x)$) does not contain an off-shell spacelike momentum (p_μ) satisfying $p_\nu p^\nu < 0$, then $\varphi(x)$ is a generalized free field.*

5.2 On the Prerequisite Knowledge

For the sake of getting new insight on off-shell quantum fields, we have to explain diversified novel notions. So, the prerequisite knowledge for understanding them necessarily takes on the character of cross-sectorial concepts. Recall that, except perhaps for the notion of quantum spin, we can safely say that the physical terminology we employ for describing a given quantum system is already used in the classical physics. In this respect, we think that the classical field theories, especially, fluid dynamics for perfect gas flow and Maxwell theory of electromagnetism, are quite informative sources of knowledge for understanding quantum fields. Actually, in order to understand our new *off-shell notions* of dark energy and dark matter, the acquaintance of

(i) Hamiltonian (H) structure of perfect fluid,
(ii) Unique invariant of generalized H structure called Ertel's potential vorticity,
(iii) So far unknown structure of Maxwell's electromagnetic theory,

seems to be necessary minimal knowledge. Therefore, in the subsequent Chap. 6, we will explain item (i) first. New notions of dark energy relating to item (iii) and dark matter to item (ii) are to be described in detail, respectively, in Chaps. 7 and 8. Based on these outcomes, in Chaps. 9 and 10, we will explain novel cosmology opened up by off-shell science and some of intriguing predictions on theoretical physics.

References

1. Nakanishi, N., Ojima, I.: Covariant Operator Formalism of Gauge Theories and Quantum Gravity. World Scientific, Singapore (1990)
2. Ojima, I.: Nakanishi-Lautrup B-Field, Crossed Product & Duality. RIMS Kokyuroku **2006**(1524), 29–37 (2006)
3. Einstein, A., Podolsky, B., Rosen, N.: Can quantum-mechanical description of reality be considered complete? Phys. Rev. **47**, 777 (1935)
4. Ojima, I.: Micro-Macro duality and emergence of macroscopic levels. Quant. Probab. White Noise Anal. **21**, 217–228 (2008)
5. Dell's Antonio, G.F.: Support of a field in p space. J. Math. Phys. **2**, 759–766 (1961)

Chapter 6
Brief Review on Generalized Hamiltonian Structure

Abstract For many of important nondissipative dynamical systems, we can say that Hamiltonian structure of the systems plays the important role of *unified theory*. On the other hand, in theoretical physics, the term *unified theory* is also used for the unification of four forces in nature, namely, (1) strong, (2) electromagnetic, (3) weak and (4) gravitational forces. As of now, dark energy and dark matter whose combined abundance ratio in the universe exceeds 90% still remains to be hidden in a veil of mystery. As you will see, a novel *off-shell science approach* we are going to employ in our present studies cast a new light on these big mysteries in theoretical physics and the Hamiltonian structure of the classical fluid dynamical field plays a key role in it. So that, this Chap. 6 is reserved for recapitulating the basic knowledge on the Hamiltonian structure.

6.1 Hamiltonian Structure of Particle and Perfect Fluid Systems

Let us start from recapitulating the essential aspects of the well-known Hamiltonian structure of a dynamical system consisting of N particles whose generalized positions and corresponding momenta, called canonically conjugate variables, are respectively denoted by (q^μ, p_μ) where $1 \leq \mu \leq N$. In such a dynamical system, any function u^μ including Hamiltonian H defined as the total energy of the system can be expressed in terms of (q^μ, p_μ). The temporal evolution of u^μ field can be expressed in a general form as

$$\dot{u}^\mu = F^\mu(u^1, ..., u^N), \quad \mu = 1, ..., N, \tag{6.1}$$

where $\dot{u}^\mu$ and F^μ denote the time derivative of u^μ and nonlinear mapping operators defined on the domain of definition called Poisson manifold. If (6.1) can be rewritten as

$$\dot{u}^\mu = \{u^\mu, H\}, \tag{6.2}$$

where $\{,\}$ denotes Poisson bracket operation having the following properties:

© The Author(s), under exclusive license to Springer Nature Switzerland AG 2025
M. Ohtsu and H. Sakuma, *Dressed Photons to Revolutionize Modern Physics*,
Nano-Optics and Nanophotonics, https://doi.org/10.1007/978-3-031-77944-2_6

$$\{f, g\} = -\{g, f\}, \quad \{f, \{g, h\}\} + \{g, \{h, f\}\} + \{h, \{f, g\}\} = 0, \qquad (6.3)$$

$$\{f, g(u^1, ..., u^N)\} = \sum_{\nu=1}^{N} \{f, u^\nu\} \frac{\partial g}{\partial u^\nu}, \qquad (6.4)$$

then, we say that the given system possesses an H structure. Notice that the above properties of the Poisson bracket are the classical analogs of the well-known relations for commutators in quantum mechanics and that (6.2) can be explicitly rewritten, in terms of canonically conjugate variables (q^μ, p_μ) as

$$\dot{p}_\mu = -\frac{\partial H}{\partial q^\mu}, \quad \dot{q}^\mu = \frac{\partial H}{\partial p_\mu}. \qquad (6.5)$$

Historically, the study of H structure of a given dynamical system was carried through by employing q^μ and p_μ. However, from a viewpoint of practical physical applications, the use of such canonical variables is not advantageous because, in not a few cases, they are either unphysical or redundant. In fact, we can readily notice that the widely employed physical variables in Eulerian representation of fluid mechanics, namely, three dimensional (3D) velocity vector $\vec{v}$ together with a couple of thermodynamical variables, for instance, pressure p and temperature T, are not canonically conjugate variables. This simple fact shows that a conventional symplectic structure generated by canonical variables is not an essential factor for H structures.

Non-canonical formulations on H structure of various fluid systems were studied notably by Morrison and Green [1, 2] who focused on the important relation between the existence of Casimir functionals and the singularity of the Poisson bracket. Since the dynamical dimension of fluid mechanical systems are infinite, for such systems, finite-dimensional representation of H formulation (6.2) must be altered accordingly. However, the essential aspect of non-canonical formulation can be readily explained by the finite-dimensional system (6.2), so that, here, we give a lucid explanation by Littlejohn [3].

For any $f(u^1, ..., u^N)$ and $g(u^1, ..., u^N)$, using (6.4) and introducing the Poisson tensor $J^{\mu\nu}(u) := \{u^\mu, u^\nu\}$, we have

$$\{f, g\} = \sum_{\mu, \nu}^{N} \frac{\partial f}{\partial u^\mu} J^{\mu\nu} \frac{\partial g}{\partial u^\nu}. \qquad (6.6)$$

Combining (6.2) and (6.6), (6.2) can be rewritten as

$$\dot{u}^\mu = \sum_{\nu=1}^{N} J^{\mu\nu} \frac{\partial H}{\partial u^\nu}. \qquad (6.7)$$

Notice that if u^μ are canonical coordinates: $(u^1, ..., u^N) = (q^1, ..., q^L; p_1, ..., p_L)$, where we have $N = 2L$, then $J^{\mu\nu}$ assumes the following form

6.1 Hamiltonian Structure of Particle and Perfect Fluid Systems

$$J^{\mu\nu} = \begin{pmatrix} 0 & I \\ -I & 0 \end{pmatrix}, \tag{6.8}$$

where 0 and I respectively denote $L \times L$ zero and unit matrices with $det(J^{\mu\nu}) = 1$.

For non-canonical variables, the Poisson tensor $J^{\mu\nu}$ becomes singular and there exist poisitive integer numbers R and K called the rank and the corank of $J^{\mu\nu}$ satisfying $N = R + K$. Under such a circumstance, there exist K linearly independent covariant vectors $\gamma_\nu^{(k)}$ satisfying

$$\sum_{\nu=1}^{N} J^{\mu\nu} \gamma_\nu^{(k)} = 0, \quad k = 1, ..., K. \tag{6.9}$$

Furthermore, if there exist $C^{(k)}$ such that $\gamma_\nu^{(k)} = \partial C^{(k)} / \partial u^\nu$, then, usuing (6.6), for any f, we obtain

$$\{f, C^{(k)}\} = 0, \quad k = 1, ..., K. \tag{6.10}$$

Since $C^{(k)}$ commutes with any function, we call it a *Casimir* function (or functional in the case of an infinite-dimensional sysytem).

The important consequence of this is that the Hamiltonian H is not unique, which can be readily seen from

$$\{u^\mu, H + \sum \lambda_{(k)} C^{(k)}\} = \dot{u}^\mu + \sum \lambda_{(k)} \{u^\mu, C^{(k)}\} = \dot{u}^\mu, \tag{6.11}$$

and the extremal points of H may change under this transformation. Directly from (6.10), we see that Casimir function C is a constant of motion satisfying

$$\dot{C} = \{C, H\} = 0. \tag{6.12}$$

For an ideal gas flow as an infinite-dimensional dynamical system, the total energy E and Casimir C [4] assume the following forms:

$$E = \int (dx^1 dx^2 dx^3) \left(\frac{\rho}{2} \vec{v} \cdot \vec{v} + \rho e(\rho, \eta) \right) \tag{6.13}$$

$$C = \int (dx^1 dx^2 dx^3) \rho F(\eta, Q), \quad Q := \frac{1}{\rho} (\nabla \times \vec{v}) \cdot \nabla \eta, \tag{6.14}$$

where ρ is gas density, e and η respectively denote the internal energy and entropy density, while Q in (6.14) is Ertel's potential vorticity which is a quite important conserved quantity in geophysical fluid mechanics and F denotes an arbitrary function to be specified under a given dynamical situation. By the property shown in (6.11), the sum of the total energy E in (6.13) and the associated Casimr C in (6.14) is called the generalized Hamiltonian of a given fluid dynamical system. We will see in Chap. 8 that the Casimir given above plays an important role in identifying dark matter field in our universe.

6.2 On Clebsch Parameterization of Barotropic Fluid

As to the Hamiltonian structure of fluid system, another quite important element for our new study is Clebsch transformation [5] or parameterization of barotropic fluid velocity v_μ. For simplicity, here, we explain the point of it in the case of three dimensional (3D) non-relativisitc fluid motion field v_μ. The original definition of Clebsch parameterization (CP) for v_μ assumes the form:

$$v_\mu = -\frac{\partial \delta}{\partial x^\mu} + \lambda \frac{\partial \phi}{\partial x^\mu}, \quad 1 \le \mu \le 3, \tag{6.15}$$

where the first and the second terms on r.h.s. respectively represent irrotational and rotational parts of velocity field v_μ.

Now, let us consider the equation of motion of a given fluid motion field in which an infinitesimally small portion of the fluid behaves like a "particle" under the influence of pressure gradient force. Applying Newton's equation of motion to such a particle, we have

$$\rho \frac{D v_\mu}{D t} = -\frac{\partial p}{\partial x^\mu}, \tag{6.16}$$

where ρ and p denote density and pressure field and we employ Cartesian coordinates system (x^1, x^2, x^3). Acceleration field $D v_\mu / D t$ is called Lagrange derivative of velocity field v_μ and its Eulerian representation takes the form of

$$\frac{D v_\mu}{D t} = \frac{\partial v_\mu}{\partial t} + v^\nu \frac{\partial v_\mu}{\partial x^\nu}. \tag{6.17}$$

In fluid mechanics, the rotational freedom of a given velocity field v_μ is represented by vorticity tensor $\omega_{\mu\nu}$ defined by

$$\omega_{\mu\nu} := \frac{\partial v_\nu}{\partial x^\mu} - \frac{\partial v_\mu}{\partial x^\nu}. \tag{6.18}$$

Using the above explicit representation of $\omega_{\mu\nu}$, (6.16) is rewritten as

$$\frac{\partial v_1}{\partial t} + \frac{\partial K}{\partial x^1} - v^2 \omega_{12} + v^3 \omega_{31} = -\frac{1}{\rho} \frac{\partial p}{\partial x^1}, \tag{6.19}$$

$$\frac{\partial v_2}{\partial t} + \frac{\partial K}{\partial x^2} - v^3 \omega_{23} + v^1 \omega_{12} = -\frac{1}{\rho} \frac{\partial p}{\partial x^2}, \tag{6.20}$$

$$\frac{\partial v_3}{\partial t} + \frac{\partial K}{\partial x^3} - v^1 \omega_{31} + v^2 \omega_{23} = -\frac{1}{\rho} \frac{\partial p}{\partial x^3}, \tag{6.21}$$

where $K := v_\nu v^\nu / 2$.

As the second step of deriving the Hamiltonian structure for the above hydrodynamical system, we are going to transform Eqs.(6.19–6.21) into the ones expressed

6.2 On Clebsch Parameterization of Barotropic Fluid

in a new coordinates system: $(t; \xi^\mu; 1 \leq \mu \leq 3)$ where ξ^μ denotes a general (or arbitrary) coordinate. Since a gradient vector $\partial\alpha/\partial\xi^\mu$ is to be transformed into $\partial\alpha/\partial x^\mu$ as a covariant vector, we have

$$
\begin{pmatrix} \frac{\partial\alpha}{\partial\xi^1} \\ \frac{\partial\alpha}{\partial\xi^2} \\ \frac{\partial\alpha}{\partial\xi^3} \end{pmatrix} = \begin{pmatrix} \frac{\partial x^1}{\partial\xi^1} & \frac{\partial x^2}{\partial\xi^1} & \frac{\partial x^3}{\partial\xi^1} \\ \frac{\partial x^1}{\partial\xi^2} & \frac{\partial x^2}{\partial\xi^2} & \frac{\partial x^3}{\partial\xi^2} \\ \frac{\partial x^1}{\partial\xi^3} & \frac{\partial x^2}{\partial\xi^3} & \frac{\partial x^3}{\partial\xi^3} \end{pmatrix} \begin{pmatrix} \frac{\partial\alpha}{\partial x^1} \\ \frac{\partial\alpha}{\partial x^2} \\ \frac{\partial\alpha}{\partial x^3} \end{pmatrix}.
\tag{6.22}
$$

Using (6.15) of CP, we obtain

$$
\frac{\partial v_\mu}{\partial t} = \left(-\frac{\partial^2\delta}{\partial t\partial x^\mu} + \lambda\frac{\partial^2\phi}{\partial t\partial x^\mu} \right) + \frac{\partial\lambda}{\partial t}\frac{\partial\phi}{\partial x^\mu},
\tag{6.23}
$$

and

$$
\omega_{\mu\nu} = \frac{\partial\lambda}{\partial x^\mu}\frac{\partial\phi}{\partial x^\nu} - \frac{\partial\lambda}{\partial x^\nu}\frac{\partial\phi}{\partial x^\mu}.
\tag{6.24}
$$

Substituting Eqs.(6.23, 6.24) into Eqs.(6.19-6.21), we get

$$
\frac{\partial}{\partial x^\mu}\left(-\frac{\partial\delta}{\partial t} + \lambda\frac{\partial\phi}{\partial t} + K \right) - \frac{\partial\lambda}{\partial x^\mu}\frac{D\phi}{Dt} + \frac{\partial\phi}{\partial x^\mu}\frac{D\lambda}{Dt} = -\frac{1}{\rho}\frac{\partial p}{\partial x^\mu}.
\tag{6.25}
$$

At this point, we introduce an important notion to distinguish different physical properties of a given fluid. A given fluid is called *barotropic* if density ρ is either constant or a certain function of pressure p, otherwise, it is called *baroclinic*. In barotropic fluid, the pressure gradient force on the r.h.s. of (6.25) can be rewritten as a potential force of the form:

$$
\frac{1}{\rho}\frac{\partial p}{\partial x^\mu} = \frac{\partial\pi}{\partial x^\mu}.
\tag{6.26}
$$

Here, we give a couple of well-known examples of barotropic fluid motions. The simplest one is the flow of incompressible fluid $\rho = \rho_0 = constant$. For such a fluid, so-called "continuity equation" representing the local mass conservation of the fluid

$$
\frac{\partial\rho}{\partial t} + \frac{(\partial\rho v^\nu)}{\partial x^\nu} = 0,
\tag{6.27}
$$

makes its fluid motion divergence free, that is to say,

$$
\frac{\partial v^\nu}{\partial x^\nu} = 0.
\tag{6.28}
$$

Another example is the isentropic flow of an ideal gas. We first note that the pressure gradient force can be rewritten as

$$-\frac{1}{\rho}\frac{\partial p}{\partial x^\mu} = T\frac{\partial s}{\partial x^\mu} - \frac{\partial h}{\partial x^\mu}, \tag{6.29}$$

where T, s and h respectively denote absolute temperature, specific entropy and enthalpy of the fluid. We see that for isentropic flow of $s = constant$, the pressure gradient force becomes a potential one.

Thus, for barotropic fluid motions, (6.25) can be rewrittens as

$$-\frac{\partial\lambda}{\partial x^\mu}\frac{D\phi}{Dt} + \frac{\partial\phi}{\partial x^\mu}\frac{D\lambda}{Dt} = -\frac{\partial H}{\partial x^\mu}, \quad H := \pi - \frac{\partial\delta}{\partial t} + \lambda\frac{\partial\phi}{\partial t} + K. \tag{6.30}$$

Using (6.22) showing the transformation relation between gradient vector $\partial\alpha/\partial x^\mu$ and $\partial\alpha/\partial\xi^\mu$, we see that (6.30) can be rewritten first as

$$
\begin{pmatrix}
\frac{\partial x^1}{\partial\xi^1} & \frac{\partial x^2}{\partial\xi^1} & \frac{\partial x^3}{\partial\xi^1} \\[2mm]
\frac{\partial x^1}{\partial\xi^2} & \frac{\partial x^2}{\partial\xi^2} & \frac{\partial x^3}{\partial\xi^2} \\[2mm]
\frac{\partial x^1}{\partial\xi^3} & \frac{\partial x^2}{\partial\xi^3} & \frac{\partial x^3}{\partial\xi^3}
\end{pmatrix}
\begin{pmatrix}
-\frac{\partial\lambda}{\partial x^1}\frac{D\phi}{Dt} + \frac{\partial\phi}{\partial x^1}\frac{D\lambda}{Dt} \\[2mm]
-\frac{\partial\lambda}{\partial x^2}\frac{D\phi}{Dt} + \frac{\partial\phi}{\partial x^2}\frac{D\lambda}{Dt} \\[2mm]
-\frac{\partial\lambda}{\partial x^3}\frac{D\phi}{Dt} + \frac{\partial\phi}{\partial x^3}\frac{D\lambda}{Dt}
\end{pmatrix}
=
\begin{pmatrix}
-\frac{\partial H}{\partial\xi^1} \\[2mm]
-\frac{\partial H}{\partial\xi^2} \\[2mm]
-\frac{\partial H}{\partial\xi^3}
\end{pmatrix}. \tag{6.31}
$$

Since (ξ^μ) is arbitrary, we can set it such that $\xi^1 = \lambda$, $\xi^2 = \phi$ and $\xi^3 = \delta$ in (6.15). With this choice, (6.31) finally becomes

$$\frac{D\phi}{Dt} = \frac{\partial H}{\partial\lambda}, \quad \frac{D\lambda}{Dt} = -\frac{\partial H}{\partial\phi}, \quad 0 = \frac{\partial H}{\partial\delta}. \tag{6.32}$$

Therefore, we have

$$\frac{D\lambda}{Dt} = -\frac{\partial H(\lambda,\phi)}{\partial\phi}, \quad \frac{D\phi}{Dt} = \frac{\partial H(\lambda,\phi)}{\partial\lambda}, \tag{6.33}$$

which is the canonical equation of motion isomorphic to (6.5).

References

1. Morrison, P.J., Greene, J.M.: Noncanonical Hamiltonian density formulation of hydrodynamics and ideal magnetohydrodynamics. Phys. Rev. Lett. **45**, 790–794 (1980)
2. Morrison, P.J., Greene, J.M.: Noncanonical Hamiltonian density formulation of hydrodynamics and ideal magnetohydrodynamics addition. Phys. Rev. Lett. **48**, 569 (1982)

References

3. Littlejohn, R.G.: Singular Poisson tensors. In: Mathematical Methods in Hydrodynamics and Integrability in Dynamical Systems (La Jolla, CA); In: AIP Conference Proceedings, vol. 88, pp. 47–66. American Institute of Physics (1982)
4. Kuroda, Y.: Symmetries and Casimir invariants for perfect fluid. Fluid Dynam. Res. **5**, 273–287 (1990)
5. Lamb, S.H.: Hydrodynamics, 6th edn. Cambridge University Press, Cambridge, UK (1930)

Chapter 7
Off-Shell Electromagnetic Field

Abstract Recall that the revolutionary reform of the classical physics done by Einstein's theory of special relativity was brought about by the recognition that the structure of physical space-time is tightly bound with the dynamics of electromagnetic waves. Prior to the revolution, it was tacitly understood that *space* and *time* are not physical entities but mathematical ones with which we can describe the structure and the time evolution of a given physical system under consideration. As we have already explained in subchapter 5.1, the core concept in our new *off-shell science* is spacelike momentum field p^μ which plays a key role in nonlinear quantum field interactions. Intuitively speaking, *space-time* can be considered as a *field* in which quantum field interactions occur, so that it would be reasonable to assume that p^μ field plays the role of spacelike part of physical space-time. In this chapter, we can show that it is actually the case. After showing this, we are going to explain the unique characteristics of de Sitter space closely related to dark energy field.

7.1 Brief Review on Free Electromagnetic Field

In the preceding chapter, we have explained the Hamiltonian structure of a barotropic fluid of which vortical motions satisfies Lagrange's vortex theorem claiming that 3D vorticity vector is advected by a co-moving fluid particle. A light field as the propagation of electromagnetic field $F_{\mu\nu}$, defined by a couple of electric $\mathbf{E}$ and magnetic $\mathbf{B}$ vectors below

$$F_{\mu\nu} := \begin{pmatrix} 0 & E^1 & E^2 & E^3 \\ -E^1 & 0 & -B^3 & B^2 \\ -E^2 & B^3 & 0 & -B^1 \\ -E^3 & -B^2 & B^1 & 0 \end{pmatrix}, \tag{7.1}$$

is mathematically regarded as the propagation of a vortex field $(\partial_\mu A_\nu - \partial_\nu A_\mu)$ along the light ray, where A_μ denotes 4D electromagnetic vector potential. That is to say, the mathematical structure of free propagation of light field is similar to that of a barotropic flow field. The reason why we are going to introduce a new notion of

© The Author(s), under exclusive license to Springer Nature Switzerland AG 2025
M. Ohtsu and H. Sakuma, *Dressed Photons to Revolutionize Modern Physics*,
Nano-Optics and Nanophotonics, https://doi.org/10.1007/978-3-031-77944-2_7

Clebsch dual field to explain the extended form of free electromagnetic field lies in this fact.

So, let us begin by reviewing the basics of free electromagnetic field propagations. A skew symmetric second-rank tensor $X_{\mu\nu} = -X_{\nu\mu}$ is often referred to as a *bivector*. So, electromagnetic field $F_{\mu\nu} = \partial_\mu A_\nu - \partial_\nu A_\mu$ is a bivector. A bivector $X_{\mu\nu}$ is said to be simple if it can be written as the exterior product of two vectors U_μ and V_ν such that $X_{\mu\nu} = 2U_{[\mu} V_{\nu]}$ where the square bracket denotes anti-symmetrization. We will see in the following subchapter that Clebsch dual spacelike electromagnetic field is represented as a simple bivector field, which is closely related to Pauli exclusion principle applied to Majorana fermionic field.

Geometrically speaking, light ray of electromagnetic field is given by the field of geodesics in 4D pseudo Riemannian manifold $\mathcal{M}$ with metric tensor $g_{\mu\nu}$. The equation of a geodesic assumes the form:

$$(\nabla_U U)_\mu := U^\nu \nabla_\nu U_\mu = U^\nu (\nabla_\nu U_\mu - \nabla_\mu U_\nu) + \nabla_\mu (U^\nu U_\nu / 2) = 0, \qquad (7.2)$$

where ∇_X denotes the covariant derivative along a vector field $X = X^\nu \partial_\nu$ associated with the Levi-Civita connection. In case that U^μ denotes relativistic four velocity vector of a given fluid particle defined as

$$U^\mu := \frac{Dx^\mu}{Ds}, \quad (Ds)^2 := g_{\mu\nu} dx^\mu dx^\nu, \qquad (7.3)$$

where x^μ are Lagrangian coordinates of a fluid particle and $(Ds)^2$ is the square of infinitesimal distance measured on the trajectory of the particle, the magnitude of U^μ can be normalized [1] as

$$g_{\mu\nu} U^\mu U^\nu = U_\nu U^\nu = 1. \qquad (7.4)$$

Notice that a fluid particle is generally accelerated by pressure gradient force F_p. If $F_p = 0$, then the particle moves with a certain constant velocity and (7.2) describes such a situation. Unlike a fluid particle trajectory, the light ray trajectory given by (7.2) is called a *null geodesic* since the magnitude of U^μ and Ds satisfy the following null conditions

$$U_\nu U^\nu = 0, \quad (Ds)^2 = g_{\mu\nu} dx^\mu dx^\nu = 0, \qquad (7.5)$$

because the magnitude of energy density of the light field is equal to that of momentum density, that is, $U_\nu U^\nu = (U^0)^2 - [(U^1)^2 + (U^2)^2 + (U^3)^2] = 0$.

Now consider Maxwell equation having the form of

$$j^\mu = \nabla_\nu F^{\nu\mu} = -g^{\mu\sigma} \nabla_\sigma (\nabla_\tau A^\tau) + [g^{\sigma\tau} \nabla_\sigma \nabla_\tau A^\mu + R^\mu{}_\sigma A^\sigma]. \qquad (7.6)$$

where $R^\mu{}_\sigma$ denotes Ricci curvature tensor usually neglected except for cosmological arguments. Such being the case, in what follows, we neglect this curvature term. A widespread view on electromagnetism is that the irrotational part of vector potential

7.1 Brief Review on Free Electromagnetic Field

A^μ is not a physical quantity, so that we can impose an arbitrary gauge condition on it. In the vacuum where we have no electric current: $j^\mu = 0$, the following is a permissible set of solutions:

$$g^{\sigma\tau}\nabla_\sigma\nabla_\tau A^\mu = 0, \tag{7.7}$$

$$g^{\mu\sigma}\nabla_\sigma(\nabla_\tau A^\tau) = 0, \tag{7.8}$$

where *non-physical* (7.8) corresponds to the well-known Lorentz gauge condition of the form: $\nabla_\tau A^\tau = 0$.

In search of a new idea on extending the well-known free electromagnetic field into the spacelike momentum domain, let us explore the possibility that the quantity

$$\phi := \nabla_\tau A^\tau \tag{7.9}$$

gives a physical field whose non-excitation state corresponds to the above-mentioned Lorentz gauge condition. Physically, this non-zero ϕ mode can be interpreted as the longitudinal mode of electromagnetic waves which is regarded as non-physical quantity in the conventional quantum electrodynamics (QED). The reason why it is non-physical is because, unlike the transverse modes of electromagnetic waves, it cannot be quantized properly. *However, classically, the existence of such a free longitudinal mode was reported both theoretically as well as experimentally* [2], *which underpins our effort on the extension of electromagnetic field into spacelike (momentum) domains.*

Once we accept the physicality of $\nabla_\nu A^\nu$, it must be treated on equal footing with $F^{\mu\nu}$ in the Lagrangian formulation of free electromagnetic field. Consider a Lagrangian density $\mathcal{L}^*$ of the form:

$$\mathcal{L}^* = \mathcal{L} + \mathcal{L}_{GF} = -\frac{1}{4}F_{\mu\nu}F^{\mu\nu} - \frac{1}{2}(\nabla_\nu A^\nu)^2, \tag{7.10}$$

where $\mathcal{L}$ and $\mathcal{L}_{GF}$ respectively denote gauge-invariant and gauge-fixing Lagrangian densities. Vanishing the first variation of $\mathcal{L}^*$, we obtain

$$(\nabla_\mu F^{\mu\nu} + g^{\nu\mu}\nabla_\mu\phi)\delta A_\nu = 0, \tag{7.11}$$

which becomes equal to (7.6) under the condition (7.7). In Nakanishi-Lautrup (NL) formalism [3] for gauge field quantization, $\nabla_\nu A^\nu$ plays a key role for the equation of propagator to have meanigful solutions. In NL formalism, a gauge-fixing Lagrangian $\mathcal{L}_{GF}$ is given by

$$\mathcal{L}_{GF} = B\nabla_\nu A^\nu + \frac{\alpha}{2}B^2, \tag{7.12}$$

where B field satisfies:

$$\nabla_\nu A^\nu + \alpha B = 0, \quad g^{\mu\nu}\nabla_\mu\nabla_\nu B = 0. \tag{7.13}$$

Comparing (7.10) with (7.12, 7.13), we see that (7.10) corresponds to Feynman gauge of $\alpha = 1$. And (7.11) leading to

$$g^{\mu\nu}\nabla_\mu\nabla_\nu\phi = 0, \tag{7.14}$$

shows that ϕ field is a physically meaninful longitudinal electromagnetic mode.

The physical interpretation of Feynman gauge can be given by the following argument. For simplicity, let us consider the vector (7.7) in a flat Minkowski space $\{x^\mu\}$ where the operator ∇_μ is replaced by ∂_μ with $R^\mu{}_\sigma = 0$. By decomposing A_ρ into the rotational α_ρ and the irrotational $\partial_\rho\chi$ parts: $A_\rho = \alpha_\rho + \partial_\rho\chi$, (7.7) is rewritten as

$$\partial^\tau\partial_\tau\alpha_\rho + \partial_\rho(\partial_\tau A^\tau) = \partial^\tau\partial_\tau\alpha_\rho + \partial_\rho(\partial^\tau\partial_\tau\chi) = 0. \tag{7.15}$$

Note that, for a given non-zero $\partial_\tau A^\tau$, the solution α_ρ to (7.15) consists of a homogeneous solution satisfying $\partial^\tau\partial_\tau\alpha_\rho^{(h)} = 0$ and inhomogeneous one $\alpha_\rho^{(i)}$ which satisfies the following equation

$$\partial^\tau\partial_\tau\alpha_\rho^{(i)} + \partial_\rho(\partial_\tau A^\tau) = 0. \tag{7.16}$$

Since $\alpha_\rho^{(h)}$ is a well-known solution of transverse modes, what we are actually interested in is $\alpha_\rho^{(i)}$. (7.16) can be regarded as a balance equation between the rotational vector α_ρ and the irrotational vector $\partial_\rho(\partial_\tau A^\tau)$. A well-known example of such a balance equation is two dimensional irrotational motion of an incompressible fluid, of which Cartesian velocity field is given be (u, v). The incompressiblity of fluid means that its motion is non-divergent, that is, $\partial_x u + \partial_y v = 0$, while the vorticity field must vanish since its motion is irrotational: $\partial_x v - \partial_y u = 0$. Therefore, such a velocity field (u, v) must have a dual representation of the form:

$$u = \partial_x\chi = \partial_y\psi, \quad v = \partial_y\chi = -\partial_x\psi, \tag{7.17}$$

where χ and ψ are the velocity potential and the streamfunction of a given velocity field (u, v). Note that (7.17) is equivalent to the Cauchy-Riemann equations in complex analysis, that is to say, χ and ψ are respectively the real and imaginary parts of a certain analytic function satisfying

$$\nabla^2\chi = 0, \quad \nabla^2\psi = 0. \tag{7.18}$$

The fact that ϕ-mode is a longitudinally propagating mode is shown directly by the following energy-momentum conservation of a given electromagnetic field $F_{\mu\nu}$ which assumes the form:

$$\partial_\nu T_\mu{}^\nu = F_{\mu\nu}(\partial^\nu\phi) = 0, \quad \Rightarrow \quad (F_{\mu\nu} \perp \partial^\nu\phi), \tag{7.19}$$

$$T_\mu{}^\nu := -F_{\mu\sigma}F^{\nu\sigma} + \frac{1}{4}\eta_\mu{}^\nu F_{\sigma\tau}F^{\sigma\tau}, \tag{7.20}$$

7.2 Clebsch Dual Electromagnetic Field

where $T_\mu{}^\nu$ and $\eta_\mu{}^\nu$ respectively denote the energy-momentum and Lorentzian metric tensors.

7.2 Clebsch Dual Electromagnetic Field

As is shown in (7.1), mathematically, electromagnetic wave field $F_{\mu\nu}$ is considered to be a relativistic vortical field expressed in terms of skew-symmetric second rank tensor: $\omega_{\nu\mu} = -\omega_{\mu\nu}$, so that it is helpful to look into certain properties of the relativistic dynamical form of a perfect fluid system. Non-relativistic equations of motion of such a system were already given in (6.19–6.21). Notice that the representation of such equations of motion is not unique and the feature of (6.19–6.21) is that they are expressed in terms of vorticity field $\omega_{\mu\nu}$. The corresponding relativistic version of (6.19–6.21) was derived by Lichnerowicz [4] and it turns out to be

$$\omega_{\mu\nu}u^\nu = T\nabla_\mu(\sigma/n), \quad \omega_{\mu\nu} := \nabla_\mu[(w/n)u_\nu] - \nabla_\nu[(w/n)u_\mu], \tag{7.21}$$

where n denotes "fluid particle number" corresponding to density ρ in the non-relativistic case, T, σ/n and w/n are respectively absolute temperature, the specific entropy and enthalpy.

In order to compare $F_{\mu\nu}$ with $\omega_{\mu\nu}$, notice first that a free Maxwell field $F_{\mu\nu}$ representing a propagating light field satisfies

$$F_{\nu\sigma}F^{\nu\sigma} = 0, \quad F_{01}F_{23} + F_{02}F_{31} + F_{03}F_{12} = 0, \tag{7.22}$$

where the first and the second equations resectively denote that the energy-momentum of this field is mass-less and the electric field (F_{01}, F_{02}, F_{03}) is perpendicular to the magnetic field (F_{23}, F_{31}, F_{12}). Now, consider a rank 4 tensor $\hat{R}_{\alpha\beta\gamma\delta}$ of the form:

$$\hat{R}_{\alpha\beta\gamma\delta} = F_{\alpha\beta}F_{\gamma\delta}, \tag{7.23}$$

which satisfies the following relations

$$\hat{R}_{\beta\alpha\gamma\delta} = -\hat{R}_{\alpha\beta\gamma\delta}, \quad \hat{R}_{\alpha\beta\delta\gamma} = -\hat{R}_{\alpha\beta\gamma\delta}, \quad \hat{R}_{\gamma\delta\alpha\beta} = \hat{R}_{\alpha\beta\gamma\delta}. \tag{7.24}$$

We see that (7.22) can be rewritten in terms of $\hat{R}_{\alpha\beta\gamma\delta}$ as

$$F_{\nu\sigma}F^{\nu\sigma} = \hat{R}_{\nu\sigma}{}^{\nu\sigma} = 0, \tag{7.25}$$

$$F_{01}F_{23} + F_{02}F_{31} + F_{03}F_{12} = \hat{R}_{0123} + \hat{R}_{0231} + \hat{R}_{0312} = 0. \tag{7.26}$$

Recall that Riemann curvature tensor $R_{\alpha\beta\gamma\delta}$ employed in the general theory of relativity satisfies

$$R_{\beta\alpha\gamma\delta} = -R_{\alpha\beta\gamma\delta}, \quad R_{\alpha\beta\delta\gamma} = -R_{\alpha\beta\gamma\delta}, \quad R_{\gamma\delta\alpha\beta} = R_{\alpha\beta\gamma\delta}, \tag{7.27}$$

$$R_\nu{}^\nu := R_{\nu\sigma}{}^{\nu\sigma}, \quad R_{\alpha\beta\gamma\delta} + R_{\alpha\gamma\delta\beta} + R_{\alpha\delta\beta\gamma} = 0. \tag{7.28}$$

We readily see first that (7.24) is identical to (7.27). As to (7.25), it corresponds to the special mass-less case of the first equation in (7.28) for which the scalar curvature of a given space-time vanishes. The second equation in (7.28) is known as the first Bianchi identity and it is equivalent to (7.26).

In relativistic fluid dynamics, if a given skew-symmetric vorticity tensor field $\omega_{\mu\nu}$ satisfies the orthogonal condition (7.26), then such a fluid motion is classified as barotropic one [5]. Directly from (7.21), we see that an isentropic ($\sigma = 0$) fluid motion is a typical example of barotropic fluid motion. In the subchapter 6.2, we have already shown that the isentropic fluid motion plays a key role for the existence of canonically conjugate Hamiltonian structure of fluid motions, which can be expressed in terms of Clebsch variables λ and ϕ. This observation motivates us to introduce CP given in (6.15) and to apply it to electromagnetic vector potential U_μ in the spacelike momentum domain, namely

$$U_\mu = \lambda \nabla_\mu \phi. \tag{7.29}$$

To appreciate the meaning of CP in applying it to the problem of electromagnetic wave prppagation, we start from the lightlike case for which we have $U_\nu U^\nu = 0$. In this case, we define λ and ϕ such that

$$\nabla^\nu \nabla_\nu \lambda - (\kappa_0)^2 \lambda = 0, \quad \nabla^\nu \nabla_\nu \phi = 0, \tag{7.30}$$

where $(\kappa_0)^{-1}$ is DP constant $l_{dp} \approx 50$ nanometer (nm) experimentally determined by Ohtsu [6] [refer also to subchapter 2.3.1 of this book]. Since a couple of gradient vectors of the canonically conjugate variables λ and ϕ play an important role in connecting the classical U_μ field to its quantum counterpart, here, we introduce an additional notations for them with an important orthogonality condition imposed on them:

$$L_\mu := \nabla_\mu \lambda, \quad C_\mu := \nabla_\mu \phi, \quad C^\nu L_\nu = 0. \tag{7.31}$$

With (7.30) and (7.31), the lightlike Clebsch dual electromagnetic field U_μ, originally given in (7.29), is rewritten as

$$U_\mu = \lambda C_\mu. \tag{7.32}$$

Utilizing the expression of geodesic given in (7.2), we can readily show that U_μ field given above satisfies a null-geodesic condition.

In conventional Maxwell's theory on electromagnetism, the field strength tensor $F_{\mu\nu}$ is expressed in terms of the curl of vector potential A_μ, namely, $F_{\mu\nu} = \nabla_\mu A_\nu - \nabla_\nu A_\mu$. Similarly, the field strength of Clebsch dual electromagnetic field $S_{\mu\nu}$ is also represented by the curl of vector potential U_μ, which turns out to be

$$S_{\mu\nu} := \nabla_\mu U_\nu - \nabla_\nu U_\mu = L_\mu C_\nu - L_\nu C_\mu. \tag{7.33}$$

7.2 Clebsch Dual Electromagnetic Field

That is to say, $S_{\mu\nu}$ can be represented as *a simple bivector field with the important orthogonal property of* $C^\nu L_\nu = 0$. The energy-momentum tensor $\hat{T}_\mu{}^\nu$ of the lightlike (trace-free) Clebsch dual electromagnetic field that can satisfies the conservation law with the following "equation of motion" for U_μ field

$$\nabla_\nu \hat{T}_\mu{}^\nu = 0, \quad U^\nu \nabla_\nu U_\mu = -S_{\mu\nu} U^\nu = 0, \tag{7.34}$$

is readily derived through exactly the same manner as the one for familiar $F_{\mu\nu}$ field, but with the following intriguing dual representations.

$$\hat{T}_\mu{}^\nu = S_{\mu\sigma} S^{\nu\sigma} = \rho C_\mu C^\nu, \quad \rho := L^\nu L_\nu < 0. \tag{7.35}$$

Since the above expression $\hat{T}_\mu{}^\nu = S_{\mu\sigma} S^{\nu\sigma}$ is identical to the one for conventional $T_\mu{}^\nu = F_{\mu\sigma} F^{\nu\sigma}$, we call this expression a wave-representation, while $\rho C_\mu C^\nu$ can be regarded as a particle representation since it is isomorphic to the energy-momentum tensor of a free particle field. Note first that, from [(7.29), (7.30) and (7.34)], we see that U_μ is a longitudinally propagating field. (7.35) says that, for such a longitudinally propagating field, particle representation $\rho C_\mu C^\nu$ becomes non-physical since the particle density ρ becomes negative. This situation corresponds to the reason why this particle mode is eliminated in conventional QED. In subchapter 5.1, we already touched on the physicality of a wave-representation in (7.35) given by Ojima's extended discussion on Nakanishi-Lautrup formalism of abelian gauge theory.

As a direct extension of the above argujments on lightlike Clebsch dual electromagnetic field $S_{\mu\nu}$, we can define spacelike $S_{\mu\nu}$ field as follows:

$$\nabla^\nu \nabla_\nu \lambda - (\kappa_0)^2 \lambda = 0, \quad L_\mu := \nabla_\mu \lambda, \tag{7.36}$$

$$\nabla^\nu \nabla_\nu \phi - (\kappa_0)^2 \phi = 0, \quad C_\mu := \nabla_\mu \phi, \tag{7.37}$$

$$C^\nu L_\nu = 0, \quad U_\mu := (\lambda C_\mu - \phi L_\mu)/2, \tag{7.38}$$

$$S_{\mu\nu} := \nabla_\mu U_\nu - \nabla_\nu U_\mu = L_\mu C_\nu - L_\nu C_\mu, \tag{7.39}$$

$$U^\nu \nabla_\nu U_\mu = -S_{\mu\nu} U^\nu + [\nabla_\mu (U^\nu U_\nu)]/2 = 0. \tag{7.40}$$

The associated energy-momentum tensor $\hat{T}_\mu{}^\nu$ becomes

$$\hat{T}_\mu{}^\nu = S_{\mu\sigma} S^{\nu\sigma} - \frac{1}{2} g_\mu{}^\nu S_{\sigma\tau} S^{\sigma\tau}. \tag{7.41}$$

Now, let us introduce a rank 4 tensor $\hat{R}_{\alpha\beta\gamma\delta}$ defined as

$$\hat{R}_{\alpha\beta\gamma\delta} := S_{\alpha\beta} S_{\gamma\delta}. \tag{7.42}$$

We can readily see that $\hat{R}_{\alpha\beta\gamma\delta}$ has exactly the same properties as the ones of Riemann curvature tensor $R_{\alpha\beta\gamma\delta}$ given in (7.27). In the theory of general relativity, Ricci curvature

$$R_\mu{}^\nu := R_{\mu\sigma}{}^{\nu\sigma}, \tag{7.43}$$

plays an significant role in interpreting the meaning of space-time curvature in terms of the energy-momentum tensor $T_\mu{}^\nu$ of a given dynamical system. That is to say, Ricci curvature $R_\mu{}^\nu$ is directly related to $T_\mu{}^\nu$ through Einstein field equation:

$$R_\mu{}^\nu - \frac{R}{2}g_\mu{}^\nu + \Lambda g_\mu{}^\nu = -\frac{8\pi G}{c^4}T_\mu{}^\nu, \quad (R := R_\nu{}^\nu), \tag{7.44}$$

where R, Λ, G and c respectively denote scalar curvature, cosmological and gravitational constants and light velocity. Sum of the first and the second terms on the l.h.s. of (7.44) is called Einstein tensor $G_\mu{}^\nu$ whose divergence vanishes identically. Since $\hat{R}_{\alpha\beta\gamma\delta}$ introduced in (7.42) has exactly the same properties as $R_{\alpha\beta\gamma\delta}$, we can define divergence-free $\hat{G}_\mu{}^\nu$ based on $\hat{R}_{\alpha\beta\gamma\delta}$. Notice that $\hat{G}_\mu{}^\nu$ thus introduced is the same as $\hat{T}_\mu{}^\nu$ defined in (7.41), which clearly shows that spacelike $S_{\mu\nu}$ field yields *physical space-time* whose existence is necessary for quantum field interactions.

7.3 Majorana Field and DP

In order to derive a quantum mechanical representation of $S_{\mu\nu}$, let us look into Dirac's equation of the from: $(i\gamma^\nu\partial_\nu + m)\Psi = 0$, which can be regarded as the square root of the timelike KG equation $(\partial^\nu\partial_\nu + m^2)\Psi = 0$. The quantum field Ψ_M that satisfies the square root of the spacelike KG equation $(\partial^\nu\partial_\nu - (\kappa_0)^2)\Psi = 0$ is shown to be the electrically neutral Majorana fermionic field $(\gamma^\nu_{(M)}\partial_\nu - \kappa_0)\Psi_{(M)} = 0$.

Owing to Pauli's exclusion principle, $\Psi_{(M)}$ with a half-integer spin of 1/2 cannot occupy the same state. A possible configuration where a couple of $\Psi_{(M)}$s form a bosonic $S_{\mu\nu}$ field can be identified using the Pauli–Lubanski 4-vector (W_μ), which describes the spin states of moving particles:

$$W_\mu = \frac{1}{2}\epsilon_{\mu\nu\lambda\sigma}M^{\nu\lambda}p^\sigma, \tag{7.45}$$

where $\epsilon_{\mu\nu\lambda\sigma}$ denotes the $4D$ totally antisymmetric Levi–Civita tensor, $M^{\nu\lambda}$ and p^σ are angular and linear momenta, respectively. As (7.45) is rewritten as

$$\begin{pmatrix} W_0 \\ W_1 \\ W_2 \\ W_3 \end{pmatrix} = \begin{pmatrix} 0 & M^{23} & M^{31} & M^{12} \\ -M^{23} & 0 & M^{03} & -M^{02} \\ -M^{31} & -M^{03} & 0 & M^{01} \\ -M^{12} & M^{02} & -M^{01} & 0 \end{pmatrix} \begin{pmatrix} p^0 \\ p^1 \\ p^2 \\ p^3 \end{pmatrix}, \tag{7.46}$$

7.3 Majorana Field and DP

we can see that two different fields, $(M^{\mu\nu}, p^\mu)$ and $(N^{\mu\nu}, q^\mu)$, which satisfy the orthogonality condition ($p^\nu q_\nu = 0$, corresponding to $C^\nu L_\nu = 0$ in (7.38)), can share the same W_μ and hence combine to form a spin 1 bosonic field, $S_{\mu\nu}$.

Having shown that how quantum Majorana fermionic field relates with the classical spacelike Clebsch dual electromagnetic field $S_{\mu\nu}$, now, we are going to look into a sketchy but essential aspect of dynamical process through which an elusive DP phenomena can be explained. As was already shown by Fig. 2.2 in subchapter 2.2, DP is created as an extremely localized field on a nano-particle. In order to construct a simple heuristic model of DP, we find that the following Aharanov et al.'s [7] pioneering study is quite informative. They investigated the behaviors of *a causal superluminal wave* when it is perturbed by a point-like disturbance of the form: $\delta(x)\delta(t)$ where $\delta(\cdot)$ denotes a δ-function, and showed that the resulting field consists of the sum of oscillatory spacelike (superluminal) modes and unstable timelike modes.

In pursueing the novel off-shell scientific studies on DP, the key ingredient is GR theorem we referred to at the end of subchapter 5.1. The reason why we pay special attention to the above Aharanov et al. study is because it deals with the behaviors of spacelike momentum fields perturbed by a point-like disturbance, which reflects the essential aspect of *a quantum light-matter field interaction generating a point-like DP field* illustrated in Fig. 2.2 (a). Their study shows that unstable timelike modes are generated through the field interactions and such unstable modes ψ_u with spherical symmetry can be shown to have a form

$$\psi_u = \exp\left[\pm k_0 x^0\right]R(r), \tag{7.47}$$

$$\frac{d^2 R}{dr^2} + \frac{2}{r}\frac{dR}{dr} - (\hat{\kappa}_0)^2 R = 0, \quad (\hat{\kappa}_0)^2 := (k_0)^2 - (\kappa_0)^2, \tag{7.48}$$

where r, k_0 and x^0 respectively denote radial coordinate, the "frequency" of a given perturbation and "time" coordinate in Lorentz-covariant expression. (7.48) shows that $R(r)$ is a timelike Yukawa potential having the following form:

$$R(r) = \frac{1}{r}\exp\left(-\frac{r}{d}\right), \quad d := (\hat{\kappa}_0)^{-1}, \tag{7.49}$$

where d represents the size of a DP field generated under a given k_0.

Since we do not have a satisfactory quantum field theory describing elusive DP phenomena, we have to adopt a suitable heuristic approach similar to the one employed in old quantum theory. In order to find such an approach, let us review first the well-known formulation of the Hamiltonian operator for a quantized harmonic oscillator. For the classical Hamiltonian of the form

$$H = \frac{p^2}{2m} + \frac{1}{2}m\omega^2 x^2, \tag{7.50}$$

where notations are conventional, the quantum basis vectors of a stationary state energy representation $|n\rangle$ satisfy

$$\hat{H}|n\rangle = e_n |n\rangle, \tag{7.51}$$

where e_n denote energy eigenvalues. Instead of using the position and momentum operators $(\hat{x}, \hat{p})$, if we use the following energy lowering $\hat{a}$ and raising $\hat{a}^\dagger$ operators defined as

$$\hat{a} := \left(\frac{\omega m}{2\hbar}\right)^{1/2} \left[\hat{x} + \left(\frac{1}{\omega m}\right) i\hat{p}\right], \quad \hat{a}^\dagger := \left(\frac{\omega m}{2\hbar}\right)^{1/2} \left[\hat{x} - \left(\frac{1}{\omega m}\right) i\hat{p}\right], \tag{7.52}$$

then, the Hamiltonian operator $\hat{H}$ in (7.51) can be rewritten as

$$\hat{H} = \left(\hat{a}^\dagger \hat{a} + \frac{1}{2}\right)\hbar\omega, \quad \epsilon_n = \left(n + \frac{1}{2}\right)\hbar\omega. \tag{7.53}$$

The operator $\hat{n} = \hat{a}^\dagger \hat{a}$ is called the number operator having the property of

$$\hat{n}|n\rangle = \hat{a}^\dagger \hat{a}|n\rangle = n|n\rangle, \tag{7.54}$$

where n denotes the number of quanta $\hbar\omega$ in the state.

Now, going back to (7.47, 7.48), let us consider the quantum mechanical implication of this classical solution by comparing the dynamical similarity between $(-k_0 x^0, +k_0 x^C)$ and $(\hat{a}, \hat{a}^\dagger)$. Since the solution $(-k_0 x^0 R(r))$ is the time-reversing one of the counterpart solution $(k_0 x^0 R(r))$, it can formally be regarded as an *antifield* of $(k_0 x^0 R(r))$ just like a pair of electron and positron. Within the framework of the classical physics, those pair fields do not interact. However, here, we set up an important working hypothesis that the pair fields, generated at the singular point, immeadiately coalesce into the localized field $R(r)$ in (7.48) just like the pair annihilation of matter vs. anti-matter. In our working hypothesis, we think that the spatial distribution of DP is given by $R(r)$ with the discretized parameter $\hat{\kappa}_0$ given by

$$(\hat{\kappa}_0)^2 := (k_0)^2 - (\kappa_0)^2, \quad k_0 = \sqrt{n}\kappa_0; \quad (n = 2, 3, 4, \cdots), \tag{7.55}$$

where the quantization $(\hat{\kappa}_0)^2 = (\kappa_0)^2 (n-1)$ is introduced as in the case of familiar photon energy quantization of $E = h\nu$. Thus DP has discrete eigen states characterized by integers n defined in (7.55).

7.4   de Sitter Space and Dark Energy

In the preceding subchapter, we have considered Majorana fermionic field whose existence is required by GR theorem referred to in subchapter 5.1. For our 4D spacetime with three spatial dimension, as can be shown from (7.46), by the restriction

7.4 de Sitter Space and Dark Energy

arising from Pauli exclusion principle, the maximum number of Majorana fermionic fields sharing the same spin vector W_μ is three, that is to say,

$$M_{\mu\nu}p^\nu = N_{\mu\nu}q^\nu = L_{\mu\nu}r^\nu = W_\mu. \qquad (7.56)$$

The state described by (7.56) is a compound state with spin 3/2 called Rarita-Schwinger state $|M3\rangle_g$. The role of *the state vector* $|M3\rangle_g$ in the sense of quantum mechanical scenario is to give the G(el'fand-)N(aimark-)S(egal) [8] cyclic vector of a mixed state which is disjoint from the vacuum state whose cyclic vector is denoted by $|0\rangle$. The important dynamical characteristics of $|M3\rangle_g$ is that Clebsch dual $S_{\mu\nu}$ vector boson field can be excited from any three pairs chosen from $[(M_{\mu\nu}p^\nu), (N_{\mu\nu}q^\nu), (L_{\mu\nu}r^\nu)]$, which propagates along one of the (x^1, x^2, x^3) directions. In view of the universality of electromagnetic field interactions, the incessant occurrence of the field interactions would make $|M3\rangle_g$ a fully occupied state for the macroscopic time-scale. In the following, we are going to show that $|M3\rangle_g$ exerts on the universe a cosmological dynamic effect identified as dark energy.

In order to apply Clebsch dual field $S_{\mu\nu}$ to cosmological problems, we first point out that, as is shown in what follows, the formulation of it derived originally for Minkowski space is readily generalized to cover the case of a curved spacetime, for which the partial derivative ∂_μ of a given field defined on the former must be replaced by the covariant derivative ∇_μ of the field defined on the latter. Such being the case, starting from Chap. 7, we have already used the symbol ∇_μ for a (covariant) derivative operator.

Actually, through the discussion covering (7.41, 7.42, 7.43, 7.44), we have shown the isomorphsm between the energy-momentum tensor of Clebsch dual field and Einstein's field equation by utilizing $\hat{R}_{\mu\nu\rho\sigma} = S_{\mu\nu}S_{\rho\sigma}$. It is clear that a curved space-time does not create any problem for defining the skew-symmetric simple bivector field $S_{\mu\nu} = L_\mu C_\nu - L_\nu C_\mu$ and hence $\hat{R}_{\mu\nu\rho\sigma} = S_{\mu\nu}S_{\rho\sigma}$. One of the notable problems we have in the case of dealing with a curved spacetime is that differential operators do not commute in general. For a given vector field V_μ on Minkowski space, we have $\partial^2_{\nu\rho}V_\mu = \partial^2_{\rho\nu}V_\mu$. On a curved space-time, however, we have $\nabla_{\nu\rho}V_\mu = \nabla_{\rho\nu}V_\mu + V_\sigma R^\sigma_{\ \mu\nu\rho}$ where $R^\sigma_{\ \mu\nu\rho}$ denotes Riemann curvature tensor, so that the order of differentiation matters. The sole exception for this non-commuting rule is the case where a vector field V_μ is represented by the gradient of a scalar field S, for which we have $\nabla_\nu S = \partial_\nu S$ and $\nabla_{\nu\rho}S = \partial^2_{\nu\rho}S - \Gamma^\sigma_{\ \nu\rho}\partial_\sigma S = \nabla_{\rho\nu}S$ because the affin connection $\Gamma^\sigma_{\ \nu\rho}$ is symmetric with respect to the subscripts ν and ρ. Notice again that the skew-symmetric Clebsch dual field $S_{\mu\nu}$ given in (7.33) is a bivector field represented in terms of the exterior product of a couple of gradient vector $L_\mu = \partial_\mu\lambda = \nabla_\mu\lambda$ and $C_\mu = \partial_\mu\phi = \nabla_\mu\phi$. Therefore, while $S_{\mu\nu}$ only contains the first derivatives of scalar fields ϕ and λ, the entire formulation of the Clebsch dual field covering, for instance, $\nabla_\nu \hat{T}_\mu^{\ \nu}$ involves the first and second derivatives of them, for the latter of which the order of differentiation does not matter.

Having stated this, we now move on to the well-known isotropic space-time structure employed in cosmological arguments:

$$ds^2 = (cdt)^2 - (R(t))^2 \left[\frac{dr^2}{1 - \xi r^2} + r^2(d\theta^2 + sin^2\theta d\phi^2) \right], \tag{7.57}$$

where ξ denotes the curvature parameter taking one of the triadic values of (0, +1, -1) and the other notations are conventional. The coordinate system employed in (7.57) is a unique co-moving (co-moving with matter) one singled out by Weyl's hypothesis on the cosmological principle with which the energy-momentum tensor $T_\mu{}^\nu$ of the universe becomes identical in form to the following one of the hydrodynamics:

$$T_\mu{}^\nu = \begin{pmatrix} \rho c^2 & 0 & 0 & 0 \\ 0 & -p & 0 & 0 \\ 0 & 0 & -p & 0 \\ 0 & 0 & 0 & -p \end{pmatrix}. \tag{7.58}$$

In addition, corresponding to (7.58), the components of metric tensor $g_{\mu\nu}$ can be chosen in such that off-diagonal elements of Einstein tensor $G_\mu{}^\nu$ are also zeros. A caveat in using this coordinate system for Clebsch dual field $S_{\mu\nu}$ is that, due to its spacelike property, the energy-momentum tensor $\hat{T}_\mu{}^\nu$ of the Clebsch dual field given by (7.41) cannot be diagonalized as in the case of (7.58) since the field resides outside the familiar timelike universe. In spite of that, the above coordinates system introduced by Weyl is a quite informative one from the viewpoint of cosmological observations, so that we think that one of the meaningful approaches to estimate the impact of $\hat{T}_\mu{}^\nu$ on our timelike universe would be to focus solely on its *diagonal components, especially the trace $\hat{T}_\nu{}^\nu$ as the sum of them* whose justification will be given shortly.

Now, we look into the energy-momentum tensor directly related to the aforementioned compound state of $|M3\rangle_g$. To avoid misunderstanding on the characters of this tensor, the following remark on fermionic fields would be helpful.

Remark

In algebraic quantum theory, the time change of a state is described by the dynamics acting on the (C*-)algebra of observables. The non-commutativity inherent to quantum theory requires the notions of quantum "observables" and "states" of a given system to be distinguished more clearly than in the classical case. Even in the classical Einstein field equation, it is true that "observables" or "physical quantities" (represented typically by the energy-momentum) and "states" (represented by the curvature of spacetime) are seen to occupy different places in a way that the former and the latter appear in the right and the left hand sides of the equation, respectively. In regard to fermionic fields, we can say that, though state changes of fermionic fields are visible, the physical quantities satisfying Fermi statistics with anti-commutation relations cannot be visible. In the conventional quantum field theory, such invisible entities as fermionic fields were introduced as an ad hoc fashion and it is not until the advent of Doplicher-Haag-Roberts theory [9] that their existence was justified through a process of reconstructing the all the members of a standard formulation of

7.4 de Sitter Space and Dark Energy

the theory involving fermionic entities, just starting from the formalism consisting of only observable data structure in the context of Galois theory.

According to the arguments in the above *remark*, the physical quantities associated with the energy-momentum tensor (7.41) derived from the spacelike Majorana fermionic field should be generally *invisible* (unobservable) in nature. The reason is as follows: the Clebsch dual field can be manipulated mathematically as if it is a classical field, similarly to the case of Schrödinger's wave equation. As far as the invisible nature of a spacelike 4 momentum vector is concerned, however, we have to take the above-mentioned property of Fermi statistics into consideration. (The close relation between the quantization of spacelike 4 momentum and Fermi statistics was pointed out, for instance, by Feinberg [10].) The key question in our analysis on dark energy is, therefore, whether we can find observable quantities or not. Since the relevant criterion for singling out such quantities may change depending on the choices of specific situations and aspects of the given problem, we have no choice but to make a good guess. The fact which seems to work as "the guiding principle" is that, within the framework of relativistic quantum field theory, any observable without exception associated with the given internal symmetry is the invariant under the action of transformation group materializing the symmetry under consideration. By extending this knowledge on the internal symmetry to the external (space-time) one, we assume that the trace $\hat{T}_{\nu}{}^{\nu}$ in (7.41) as the invariant of general coordinate transformation is observable since it is directly related to the actual observable quantity of the expansion rate of the universe through the isomorphism between $\hat{T}_{\mu}{}^{\nu}$ in (7.41) and Einstein tensor $G_{\mu}{}^{\nu} := R_{\mu}{}^{\nu} - Rg_{\mu}{}^{\nu}/2$ in (7.44). To implement our analyses on dark energy, for the sake of simplicity, we take two-stage approach I and II. In the first stage I, we confine the scope of our argument to sub-Hubble scales in which the space-time of the isotropic universe can be regarded as Minkowski space in an approximate sense. Then, in the second stage II, we smoothly extend our argument beyond those limits to cover the entire curved spacetime.

Stage I analyses

Firstly, to incorporate the fundamental quantum condition of photon energy : ($E = h\nu$) into the Clebsch dual field, let us consider first the lightlike case given by (7.35), where we have $\hat{T}_{\mu}{}^{\nu} = \rho C_{\mu}C^{\nu} = (\partial_{\sigma}\lambda\partial^{\sigma}\lambda)\partial_{\mu}\phi\partial^{\nu}\phi$. In this case, since the second equation in (7.30), namely, $\partial^{\nu}\partial_{\nu}\phi = 0$ corresponds to the lightlike dynamical factor, we may set the physical dimension of $\rho = \partial_{\sigma}\lambda\partial^{\sigma}\lambda$ such that $\rho = L_{\nu}L^{\nu}$ where the physical dimension of L_{ν} is length. The reason why we introduce such a dimensional ρ is simply because the physical dimension of $\hat{T}_{\mu}{}^{\nu}$ becomes equal to that of ϕ^2 in such a case. Using plane wave expressions of

$$\phi = \hat{\phi}_{c}\exp(ik_{\nu}x^{\nu}), k_{\nu}k^{\nu} = 0; \ \lambda = iN_{\lambda}\hat{\lambda}_{0}\exp(il_{\nu}x^{\nu}), l_{\nu}l^{\nu} = -(\kappa_{0})^{2}, \quad (7.59)$$

where i, $\hat{\lambda}_0$ and N_{λ} denote the imaginary unit, the quantized elemental amplitude that can be set as $\hat{\lambda}_0 := (\kappa_0)^{-2}$ and the number of such an elemental mode. With such a choice of $\hat{\lambda}_0$, we have

$$(C_\mu)^* C^\nu = k_\mu k^\nu \hat{\phi}_c (\hat{\phi}_c)^*, \quad \rho = (iN_\lambda)^2 (\kappa_0)^{-2}, \tag{7.60}$$

where $(\bullet)^*$ denotes the complex conjugate of $(\bullet)$.

Next, we introduce Cartesian coordinates x^1, x^2, and x^3 such that the k vector for ϕ is parallel to the x^1 direction and consider a rectangular parallelepiped V spanned by the length vector $(1/k_1, 1, 1)$. Using (7.60) and $k_0 = \nu_0/c$ where c denotes the light velocity, the volume integration of $\hat{T}_0{}^0/(iN_\lambda)^2$ over V as the energy per quantum becomes

$$\frac{1}{(iN_\lambda)^2} \int_V \hat{T}_0{}^0 dx^1 dx^2 dx^3 = (\kappa_0)^{-2} \epsilon [\hat{\phi}_c (\hat{\phi}_c)^*] \frac{\nu_0}{c} \quad \rightarrow \quad h = \frac{1}{c} (\kappa_0)^{-2} \epsilon [\hat{\phi}_c (\hat{\phi}_c)^*], \tag{7.61}$$

from which the condition corresponding to $E = h\nu$ is identified as the second equation in (7.61), where ϵ denotes a unit square meter.

For the non-lightlike (spacelike) case of $U^\nu (U_\nu)^* < 0$, using (7.41), we have $\hat{T}_\nu{}^\nu = -S_{\sigma\tau} S^{\sigma\tau}$. So, defining $|\hat{T}_\nu{}^\nu| := -S_{\sigma\tau} (S^{\sigma\tau})^*$, we obtain

$$|\hat{T}_\nu{}^\nu| := -S_{\sigma\tau} (S^{\sigma\tau})^* = 4(\kappa_0)^2 [U^\nu (U_\nu)^*] = -2(\kappa_0)^4 [\hat{\lambda}_c (\hat{\lambda}_c)^*][\hat{\phi}_c (\hat{\phi}_c)^*] < 0, \tag{7.62}$$

where $\hat{\lambda}_c$ and $\hat{\phi}_c$ denote respectively constant amplitudes of plane waves of λ and ϕ fields having the form $\psi = \hat{\psi}_c \exp[i(k_\nu x^\nu)]$. By applying the same argument given in the lightlike case, we may set $\hat{\lambda}_c = N_\lambda \hat{\lambda}_0 = N_\lambda (\kappa_0)^{-2}$. Therefore, for $N_\lambda = 1$, we have

$$|\hat{T}_\nu{}^\nu|_1 = -2[\hat{\phi}_c (\hat{\phi}_c)^*]. \tag{7.63}$$

Needless to say, Clebsch dual wave field $S_{\mu\nu}$, as in the case of an electromagnetic wave ($F_{\mu\nu} = -F_{\nu\mu}$), has a propagating direction. For isotropic radiation, we need three plane wave fields, any pair of which is mutually orthogonal. Such three fields are given, for instance, by (S_{23}, S_{02}), (S_{31}, S_{03}) and (S_{12}, S_{01}). $\hat{T}_\mu{}^\nu(3)$ derived by the superposition of these fields with $S_{23} = S_{31} = S_{12} = \sigma$ and $S_{01} = S_{02} = S_{03} = \tau$ turns out to be

$$\hat{T}_\mu{}^\nu(3) = \begin{pmatrix} -3\sigma^2 & 0 & 0 & 0 \\ 0 & \alpha & -\beta & -\beta \\ 0 & -\beta & \alpha & -\beta \\ 0 & -\beta & -\beta & \alpha, \end{pmatrix}; \begin{pmatrix} \alpha := 2\tau^2 - \sigma^2 \\ \beta := \tau^2 + \sigma^2 \end{pmatrix}, \tag{7.64}$$

which is the energy-momentum tensor of a newly proposed form of *the anti dark energy* (dark energy with negative energy density of $\hat{T}_0{}^0(3) = -3\sigma^2 < 0$) in our argument. As we will see shortly, the energy-momentum tensor of the dark energy $^*\hat{T}_\mu{}^\nu(3)$ with positive $^*\hat{T}_0{}^0(3)$ given by (7.66) *having exactly the same trace* as that of $\hat{T}_\mu{}^\nu(3)$ in (7.64) can be introduced.

7.4 de Sitter Space and Dark Energy

The reason why we consider $^*\hat{T}_0{}^0(3)$ is related to the twin structure of universe which will be explained in subchapter 9.3 in detail. The original idea of twin universes was proposed by Petit [11]. His twin universes consist of matter and anti-matter components with time-reversal symmetry that can be shown by the following classical Hamiltonian system

$$\frac{dp_\mu}{dt} = -\frac{\partial H}{\partial q^\mu}; \quad \frac{dq^\mu}{dt} = \frac{\partial H}{\partial p_\mu}, \quad \Leftrightarrow \quad \frac{dp_\mu}{d(-t)} = -\frac{\partial(-H)}{\partial q^\mu}; \quad \frac{dq^\mu}{d(-t)} = \frac{\partial(-H)}{\partial p_\mu}. \tag{7.65}$$

Namely, the time evolution of a given dynamical system with negative energy (Hamiltonian) can be reinterpreted as the backward time evolution from the view point of the counterpart system with positive energy. In quantum level, we also encounter similar reinterpretations in Feynman diagrams to distinguish the anti-particle arising from a pair creation. The existence of negative energy in quantum level is actually shown by *the two-sided (positive and negative) energy spectra of the Tomita-Takesaki extension of the thermal equilibrium* [12] of Kubo-Martin-Schwinger (KMS) state. At the macroscopic classical level, however, there is no hint of the existence of anti-matter in abundance, so that the weak energy condition (positivity of the energy) in the classical general theory of relativity related to the stability of a given dynamical system under consideration must be tied with the matter (with positive energy) dominated property of our universe.

Notice that, with the Hodge dual exchanging between (σ, τ) and $(i\tau, i\sigma)$ in (7.64), $\hat{T}_\mu{}^\nu(3)$ turns into the following $^*\hat{T}_\mu{}^\nu(3)$ with $\hat{T}_\nu{}^\nu(3) =^* \hat{T}_\nu{}^\nu(3) = -6\sigma^2 + 6\tau^2$:

$$^*\hat{T}_\mu{}^\nu(3) = \begin{pmatrix} 3\tau^2 & 0 & 0 & 0 \\ 0 & {}^*\alpha & \beta & \beta \\ 0 & \beta & {}^*\alpha & \beta \\ 0 & \beta & \beta & {}^*\alpha, \end{pmatrix}; \begin{pmatrix} {}^*\alpha := -2\sigma^2 - \tau^2 \\ \beta := \tau^2 + \sigma^2 \end{pmatrix}, \tag{7.66}$$

In $^*\hat{T}_\mu{}^\nu(3)$, the transformed 4-momentum vector density in the first row (in comparison to that in (7.64)) changes the sign, while the trace of it remains exactly the same as that of $\hat{T}_\mu{}^\nu(3)$ in (7.64). Thus, the physical meaning of the dual existence of (7.64) and (7.66) is that the notion of matter-antimatter duality can be extended to the dark energy model based on the Clebsch dual field. Notice that the diagonal components of $^*\hat{T}_\mu{}^\nu(3)$ resemble the artificial partition of the diagonal components of $\Lambda g_\mu{}^\nu$ into $\rho_\Lambda = \lambda c^4/(8\pi G)$ and $p_\Lambda = -\Lambda c^4/(8\pi G)$ (cf. (7.58)) already employed as the hypothetical equation of state of dark energy in the conventional cosmology.

Stage II analyses

Combining (7.63, 7.64) and (7.66), we get

$$^*\hat{T}_\nu{}^\nu(3) = \hat{T}_\nu{}^\nu(3) = -6\sigma^2 + 6\tau^2 = -6[\hat{\phi}_c(\hat{\phi}_c)^*]. \tag{7.67}$$

As we already pointed out, the isomorphism between $\hat{T}_\mu{}^\nu$ given in (7.41) and Einstein tensor $G_\mu{}^\nu$ in (7.44) can be extended to the one in a curved space-time. The inclusion of $\hat{T}_\nu{}^\nu(3)$ in Einstein field equation induces non-zero scalar curvature in the universe. The configuration of such a specific universe is described by a $4D$ pseudo-hypersphere $\mathfrak{D}$ with a certain "radius" $R_{dS} := \sqrt{3/\Lambda_{de}}$ embedded in a fifth dimensional Minkowski space $(ix^0, x^1, x^2, x^3, x^4)$. The pseudo-hypersphere $\mathfrak{D}$ is known as de Sitter space whose metric invariant ds^2 can be rewritten with polar coordinates (r, θ, ϕ) as

$$ds^2 = (cdt)^2 - (R_0)^2 \exp\left(2\sqrt{\frac{\Lambda_{de}}{3}}ct\right)\{dr^2 + r^2(d\theta^2 + \sin^2\theta)d\phi^2\}, \qquad (7.68)$$

where R_0 denotes a constant initial radius of our expanding universe. By comparing (7.68) with (7.57), we see that the curvatue parameter ξ of de Sitter space is zero, which shows that the analyses in the first stage I can be extended smoothly to the second stage II. Since de Sitter space is a unique solution of Einstein field equation with the cosmological term of $\Lambda_{de}g_\mu{}^\nu$, we see that the impact of $\hat{T}_\mu{}^\nu(3)$ on Einstein field equation can be observed through the cosmological constant Λ_{de}.

The observational data available to us on our expanding universe is the cosmological constant Λ_{obs} derived on the assumption that the dark energy may be modeled by the cosmological term $\Lambda g_\mu{}^\nu$. If the dark energy is modeled by $\Lambda_{obs}g_\mu{}^\nu$, then Einstein field equation with the sign convention of $R_{\mu\nu} = R^\sigma{}_{\mu\nu\sigma}$ becomes the first equation in (7.69), and if it is modeled by $*\hat{T}_\mu{}^\nu(3)$, then Einstein field equation becomes the second one in (7.69):

$$R_\mu{}^\nu - \frac{R}{2}g_\mu{}^\nu + \Lambda_{obs}g_\mu{}^\nu = -\frac{8\pi G}{c^4}T_\mu{}^\nu, \qquad R_\mu{}^\nu - \frac{R}{2}g_\mu{}^\nu = -\frac{8\pi G}{c^4}(T_\mu{}^\nu + *\hat{T}_\mu{}^\nu(3)),$$
$$(7.69)$$

which suggests that one of the meaningful observational validations of our dark energy candidate model $*\hat{T}_\mu{}^\nu(3)$ would be to compare the traces of $\Lambda_{obs}g_\mu{}^\nu$ and $(-8\pi G/c^4)*\hat{T}_\mu{}^\nu(3)$. Since the trace of $*\hat{T}_\mu{}^\nu(3)$ is the same as that of $\hat{T}_\mu{}^\nu(3)$, we see that, using (7.61, 7.67) together with the experimentally determined value of $(\kappa_0)^{-1} \approx 50$ nm [6] [refer also to subchapter 2.3.1 of this book], we can determine the magnitude of $*\hat{T}_\nu{}^\nu(3)$ and can introduce

$$-\Lambda_{de} = -\Lambda_{DP} = (8\pi G/c^4)\hat{T}_\nu{}^\nu(3)/g_\nu{}^\nu \approx -2.47 \times 10^{-53} m^{-2}, \qquad (7.70)$$

which may be regarded as the "reduced cosmological constant" of our dark energy model $*\hat{T}_\mu{}^\nu(3)$, while the value of Λ_{obs} derived by Planck satellite observations [13] is $\Lambda_{obs} \approx 3.7 \times 10^{-53} m^{-2}$. Thus, we can show that $|M3\rangle_g$ is a promising candidate for dark energy.

7.5 On the Quantization of de Sitter Space

For a plane wave solution ($\lambda = \hat{\lambda}_c \exp[i(k_\nu x^\nu)]$) to the spacelike KG equation (7.36), $L_\nu = \nabla_\nu \lambda$ satisfies

$$L^\nu(L_\nu)^* = -(\kappa_0)^2[\hat{\lambda}_c(\hat{\lambda}_c)^*] = const. < 0. \qquad (7.71)$$

As we have already explained, the vector L^μ lies on a manifold called de Sitter space, which is a pseudo-hypersphere $\mathfrak{D}$ with a certain constant radius embedded in R^5. Snyder [14] discussed the unique role of this space in space-time quantization. He showed that with the introduction of the hypothetical momentum 5-vector η^μ ($0 \leq \mu \leq 4$) in R^5 constrained to lie on the de Sitter space:

$$\eta^\nu \eta_\nu = -(R_{dS})^2 = const., \qquad (7.72)$$

the following commutation relations on momentum p_μ and position x^μ are derived. For the definitions of p_μ, $\hat{p}_\mu$ and $\hat{x}^\mu$, we have

$$p_\mu := \frac{\bar{h}}{l_p}\frac{\eta_\mu}{\eta_4}, \quad p^\nu p_\nu = \left(\frac{\bar{h}}{l_p}\right)^2\left[1 - \left(\frac{R_{dS}}{\eta_4}\right)^2\right] < 0, \qquad (7.73)$$

$$\hat{p}_\mu := -\frac{i\bar{h}}{l_p\eta_4}\frac{\partial}{\partial\eta_\mu}, \quad \hat{x}^\mu := il_p\left(\eta_4\frac{\partial}{\partial\eta_\mu} - \xi_\mu\eta_\mu\frac{\partial}{\partial\eta_4}\right); \quad (0 \leq \mu \leq 3), (7.74)$$

where l_p denotes the Planck length and ξ_μ takes a value of -1 when $\mu = 0$ and 1 when $\mu \neq 0$, from which we obtain

$$\left[\hat{x}^\mu, \hat{p}_\mu\right] = i\bar{h}\left[1 + \xi_\mu\left(\frac{l_p}{\bar{h}}\right)^2(p_\mu)^2\right], \left[\hat{x}^\mu, \hat{p}_\nu\right] = \left[\hat{x}^\nu, \hat{p}_\mu\right] = i\bar{h}\left(\frac{l_p}{\bar{h}}\right)^2 p_\mu p_\nu,$$

$$\left[\hat{x}^i, \hat{x}^j\right] = \frac{i(l_p)^2}{\bar{h}}\epsilon_{ijk}L_k, \quad \left[\hat{x}^0, \hat{x}^i\right] = \frac{i(l_p)^2}{\bar{h}}M_i; \quad 1 \leq (i, j, k) \leq 3, \qquad (7.75)$$

where ϵ_{ijk} is Edington's epsilon and L_i and M_i are angular momentum vectors generated respectively by (spatial-spatial) and (spatial-temporal) rotations. *Snyder further showed that the "Lorentz transformation" in his spacelike momentum space* $\{\eta^\mu\}$, *($0 \leq \mu \leq 3$) naturally induces the Lorentz transformation in the usual space-time* $\{x^\mu\}$. Thus, the energy-momentum tensor $\hat{T}_\mu{}^\nu$ of CD field given in (7.41) can be regarded as the one constructed on this *Snyder's momentum* "space-time" η^μ with Lorentz invariance that can be quantized. That is to say, the isomorphism between $\hat{T}_\mu{}^\nu$ and Einstein tensor $G_\mu{}^\nu = R_\mu{}^\nu - Rg_\mu{}^\nu/2$ referred to at the end of subsection 7.2 implies that the quantization of CD field through Majorana fermionic field must be closely related to Snyder's quantization of de Sitter space.

References

1. Landau, L.D., Lifshitz, E.M.: Fluid mechanics. In: Course of Tehoretical Physics, 2nd ed., Elservier, Oxford, UK, Vol. 6 (1987)
2. Cicchitelli, L., Hora, H., Postle, R.: Longitudinal field components for laser beams in vacuum. Phys. Rev. A **41**, 3727–3732 (1990)
3. Nakanishi, N., Ojima, I.: Covariant Operator Formalism of Gauge Theories and Quantum Gravity. World Scientific, Singapore (1990)
4. Lichnerowicz, A.: Relativistic Hydrodynamics and Magnetohydrodynamics. Benjamin, New York (1967)
5. Sakuma, H., Ojima, I., Saigo, H., Okamura, K.: Conserved relativistic Ertel's current generating the vortical and thermodynamic aspects of space-time. Int. J. Mod. Phys. A **37**(22), 2250155 (2022), https://doi.org/10.1142/S0217751X2250155X
6. Sakuma, H., Ojima, I., Ohtsu, M., Ochiai, H.: Off-shell quantum fields to connect dressed photons with cosmology. Symmetry **12**, 1244 (2020). https://doi.org/10.3390/sym12081244
7. Aharonov, Y., Komar, A., Susskind, L.: Superluminal behavior, causality and instability. Phys. Rev. **182**, 1400–1402 (1969)
8. Gelfand, I.M., Naimark, M.A.: On the imbedding of normed rings into the ring of operators on a Hilbert space. Matematicheskii Sbornik **12**(2), 197–217 (1943)
9. Doplicher, S., Haag, R., Roberts, J.E.: Fields, observables and gauge transformations I & II. Comm. Math. Phys. **13**, 1–23 (1969)
10. Feinberg, G.: Possibility of faster-than-light-light particles. Phys. Rev. **159**, 1089–1105 (1967)
11. Petit, J.P.: Twin Universes cosmology. Astrophys. Space Sci. **226**, 273–307 (1995)
12. Takesaki, M.: Tomita's Theory of Modular Hilbert Algebra and its Applications, Lecture Notes in Mathematics, Vol. 128. Springer (1970). https://doi.org/10.1007/BFb0065832, ISBN: 978-3-540-04917-3
13. Liu, H.: https://www.quora.com/What-is-the-best-estimate-of-the-cosmological-constant. Accessed 15 April 2020
14. Snyder, H.S.: Quantized space-time. Phys. Rev. **71**, 38 (1947)

Chapter 8
Novel Aspect of Conformal Gravity

Abstract In the previous Chap. 7, we have shown that spacelike momentum field p^μ so far ignored as *non-physical ghost field* is an inevitable constituent element of the spacelike part of physical space-time. If so, then, a natural question we will ask next would be "what about the timelike part of physical space-time ?" In order to answer this intriguing question, we first show that the Hamiltonian structure we have touched upon in Chap. 6 again plays an important role in leading the discussion. Except for a couple of towering accomplishments in modern physics, namely, quantum physics and Einstein's theory of (special & general) relativity, so-called *Nonlinear Science* emerged in 1980s with catch-phrases of "chaos", "soliton" and "fractal" provided a lot of important new concepts in modern physics. In our present context, we think that the concept of *self-similarity* in "fractal" becomes especially important. In Chap. 7, we have investigated scale-free dynamics on electromagnetic field. In geophysical fluid dynamics, the most important dynamical element is known as Ertel's potential vorticity (PV) closely related to the Casimir in the generalized Hamiltonian we discussed in Chap. 6. Owing to the pioneering study of Penrose, we know that space-time can be represented as a certain kind of spin-network, so that we conjecture that, by applying the notion of *self-similarity*, the concept of PV may be extended to the timelike part of physical space-time. The main theme of this chapter is to show that it is actually the case.

8.1 Non-relativistic Representation of Ertel's Potential Vorticity

Although the notion of Ertel's potential vorticity (PV) plays a quite important role in this chapter, its importance is not recognized at all except in the field of geophysical fluid dynamics. Such being the case, we begin with the well-known equation of motion of a perfect fluid in non-relativistic fluid mechanics:

$$D_t v_\mu := \partial_t v_\mu + v^\nu \partial_\nu v_\mu = -\frac{1}{\rho} \partial_\mu p, \tag{8.1}$$

© The Author(s), under exclusive license to Springer Nature Switzerland AG 2025
M. Ohtsu and H. Sakuma, *Dressed Photons to Revolutionize Modern Physics*,
Nano-Optics and Nanophotonics, https://doi.org/10.1007/978-3-031-77944-2_8

where the notations are conventional. A given fluid is classified as either barotropic or baroclinic based on its form of $\partial_\mu p/\rho$; that is, if $p = p(\rho)$ or $\rho = \rho_0 = const.$, the fluid is barotropic; otherwise, it is baroclinic. In other words, the barotropic or baroclinic nature of the fluid is characterized by whether $\partial_\mu p/\rho$ is a conservative force field or not, which has a decisive influence on whether the associated vorticity field is conservative or not. Thus, we refer to the *Langange's vortex theorem for barotropic flows*, which shows that vortices are free of generation and extinction.

Since we are primarily interested in vorticity and entropy fields, it is useful to rewrite the baroclinic form of (8.1) in terms of the vorticity $\zeta_{\mu\nu}$ and specific (i.e., per unit mass) entropy s fields by using the following first law of thermodynamics (8.2) and the vector identity given in (8.3):

$$- dp/\rho = T ds - dw, \quad \text{with} \quad \partial_t s + v^\nu \partial_\nu s = 0, \tag{8.2}$$

$$v^\nu \partial_\nu v_\mu = v^\nu (\partial_\nu v_\mu - \partial_\mu v_\nu) + \partial_\mu (v^\nu v_\nu/2), \tag{8.3}$$

where T and w are the absolute temperature and specific enthalpy, respectively. Using (8.2) and (8.3), (8.1) can be rewritten as

$$\partial_t v_\mu + \partial_\mu (w + v^\nu v_\nu/2) - \zeta_{\mu\nu} v^\nu = T \partial_\mu s. \tag{8.4}$$

The most well-known example of a baroclinic fluid is the atmosphere, for which the ideal gas law can be applied with a high degree of accuracy. The atmosphere is particularly important in our discussion because it provides a useful fluid dynamic system with a nonuniform entropy distribution in both the vertical and meridional directions; furthermore, energetic vortical fields known as baroclinic eddies play important dynamical roles in heat transport along the meridional direction. In the dynamics of this heat transport, there is a strong correlation between $3D$ vorticity ζ and entropy gradient ∇s vector fields, which can be described by Ertel's potential vorticity Q [1], defined as

$$Q := \frac{1}{\rho}(\zeta \cdot \nabla s), \quad \partial_t Q + v^\nu \partial_\nu Q = 0. \tag{8.5}$$

The above Q is the most important conserved quantity in the field of geophysical fluid dynamics.

It should be noted that the importance of baroclinicity in the atmosphere varies with scale. In general, in typical laboratory experiments using air, such as wind tunnel studies, air flows behave as barotropic fluids since the entropy gradient is negligible; however, for air flows with horizontal scales greater than several hundred or a few thousand kilometers, baroclinicity becomes a nonnegligible dynamical factor. In subchapter 5.1, we discussed MMD theory proposed by Ojima, which indicates that the actual world in which we live has a duality structure that bridges quantum and classical worlds in a consistent fashion. Notice that Ojima's view is decisively different from *the prevailing view that the laws of classical physics are not fundamental*

8.1 Non-relativistic Representation of Ertel's Potential Vorticity 97

but are approximated ones since "genuine fundamental laws" are those of quantum mechanics. We think that the scale dependency of baroclinicity, represented by Q in the comparison of barotropic and baroclinic flows, is analogous to that of the Planck constant h in the comparison of quantum and classical physics in MMD theory. To show that this resemblance is not superficial but rather has essential implications for the main issue in our discussion, here we are going to investigate the generalized Hamiltonian (H) structure of baroclinic flows as the extension of the theme discussed in subchapter 6.2.

The generalized H structure was derived from the so-called non-canonical form of the H formulation, of which important seminal work was initiated by Arnol'd [2] and further developed by a group of applied mathematicians and physicists [3]. For a baroclinic perfect fluid dynamics system with Eulerian representation, the generalized H, denoted by H_G, has the form [4]:

$$H_G := E + C_F, \quad E := \int \langle \rho v^\nu v_\nu / 2 + \rho e(\rho, s) \rangle dV, \quad C_F := \int \langle \rho F(s, Q) \rangle dV,$$

$$(8.6)$$

where E and C_F are the total energy with $e(\rho, s)$ being the internal energy density, a Casimir constructed by an arbitrary function F of s and Q. Since the given fluid dynamical system can be described by five independent variables, namely, v^μ, ($1 \leq \mu \leq 3$) and two thermodynamical variables, we choose ρ and s as the thermodynamic variables because Q is expressed in terms of these two variables. First, when we compare the conservative quantity C_F with the total energy E, we can see that it is not merely an additional constant of motion. This occurs because both E and C_F are "complete"in the sense that they include all five variables. In other words, they are equal pairs of "complete"constants of motion.

The significant advantage of H_G over E becomes clear when we consider the stability of a given steady state of the fluid because any given steady state of the baroclinic flow can be represented by the condition that the first variation in H_G vanishes; that is,

$$\delta H_G = \delta E + \delta C_F = 0, \quad (8.7)$$

which can be rewritten as the combination of the steady-state version of (8.4) and

$$F - Q \frac{\partial F}{\partial Q} + B(s, Q) = 0. \quad (8.8)$$

(8.8) shows that an arbitrary function F can be determined by Bernoulli's function $B(s, Q)$, which characterizes the given steady state. Then, we can demonstrate *the formal stability* of the state [5] (the stability of a given steady state for infinitesimally small amplitude perturbations with arbitrary forms) if the second variation of H_G is sign definite. In our present discussion, a particularly important aspect of (8.7) is that the balance between δE and δC_F can be regarded as a unique "interaction" between two dynamics with and without explicit forms of Q. It is clear that the phase-space trajectory of a linearly unstable mode, such as those represented by one of the separatrices of H_G, i.e., $\delta^2 H_G = 0$, cannot be described without C_F.

8.2 Converted Form of the Relativistic Equation of Motion and of PV

Let us introduce a non-dimensional four-velocity vector u^μ that satisfies the normalization condition $u^\nu u_\nu = 1$. In this subchapter, for simplicity, unless otherwise stated, we develop our arguments within the framework of special relativity. The following basic arguments on the relativistic equations from ((8.9)–(8.15)) are given by Landau & Lifshitz [1 in Sect. 7, (pp. 506–508)]. Regarding the first law of thermodynamics, we have

$$-\frac{dp}{n} = Td\left(\frac{\sigma}{n}\right) - d\left(\frac{w}{n}\right), \tag{8.9}$$

where n denotes the "particle number" corresponding to the density ρ in the non-relativistic case which satisfies a so-called *continuity equation* of the from:

$$\partial_\nu(nu^\nu) = 0, \tag{8.10}$$

and σ/n and w/n are the specific entropy and enthalpy, respectively, as in (8.2). *The assumption* (8.10) *means that we only consider low energy states of a relativistic perfect fluid.* The energy-momentum tensor for a perfect fluid has the following form:

$$T^{\mu\nu} = wu^\mu u^\nu - pg^{\mu\nu}. \tag{8.11}$$

The tensor divergence of (8.11) gives the following equations of motions:

$$\partial_\nu T_\mu{}^\nu = u_\mu \partial_\nu(wu^\nu) + wu^\nu \partial_\nu u_\mu - \partial_\mu p = 0. \tag{8.12}$$

The projection of (8.12) in the direction u^μ can be calculated by $u^\mu \partial_\nu T_\mu{}^\nu = 0$. Using (8.9), this becomes

$$u^\nu \partial_\nu(\sigma/n) = 0, \tag{8.13}$$

which corresponds to the second equation in (8.2). Next, we calculate the component of $\partial_\nu T_\mu{}^\nu = 0$ perpendicular to u^μ as

$$\partial_\nu T_\mu{}^\nu - u_\mu u^\nu \partial_\sigma T_\nu{}^\sigma = 0, \tag{8.14}$$

which yields

$$wu^\nu \partial_\nu u_\mu - \partial_\mu p + u_\mu u^\nu \partial_\nu p = 0. \tag{8.15}$$

This equation corresponds to the non-relativistic form of (8.1).

8.2 Converted Form of the Relativistic Equation of Motion and of PV 99

As we have noted, (8.4) in subchapter 8.1 is preferable to the form of (8.1), since it is expressed in terms of the vorticity and the entropy field. The derivation of the relativistic form of (8.4) is straightforward and has the following form:

$$\omega_{\mu\nu}u^\nu = T\partial_\mu(\sigma/n), \quad \omega_{\mu\nu} := \partial_\mu[(w/n)u_\nu] - \partial_\nu[(w/n)u_\mu]. \tag{8.16}$$

The details of derivation of (8.16) are given in appendix A of 5 in Sect. 7. (8.13) is included in (8.16), as shown by $0 = u^\mu\omega_{\mu\nu}u^\nu = Tu^\mu\partial_\mu(\sigma/n)$. We can also observe that (8.16) remains valid for curved spacetime if we replace ∂_μ with a covariant derivative ∇_μ. In particular, we have $\omega_{\mu\nu} = \partial_\mu v_\nu - \partial_\nu v_\mu = \nabla_\mu v_\nu - \nabla_\nu v_\mu$.

To appreciate the importance of (8.16), we use an explicit (writing down all the elements) matrix representation of (8.17) below to derive the conserved Ertel's current.

$$\omega_{\mu\nu} = \begin{pmatrix} 0 & \omega_{01} & \omega_{02} & \omega_{03} \\ -\omega_{01} & 0 & \omega_{12} & -\omega_{31} \\ -\omega_{02} & -\omega_{12} & 0 & \omega_{23} \\ -\omega_{03} & \omega_{31} & -\omega_{23} & 0 \end{pmatrix}. \tag{8.17}$$

First, after defining the pseudo-scalar Ω in (8.18), we introduce $\hat{\omega}^{\mu\nu}$, which is the Hodge dual of $\omega_{\mu\nu}$, i.e.,

$$\Omega := \omega_{01}\omega_{23} + \omega_{02}\omega_{31} + \omega_{03}\omega_{12}, \tag{8.18}$$

$$\hat{\omega}^{\mu\nu} = \begin{pmatrix} 0 & -\omega_{23} & -\omega_{31} & -\omega_{12} \\ \omega_{23} & 0 & -\omega_{03} & \omega_{02} \\ \omega_{31} & \omega_{03} & 0 & -\omega_{01} \\ \omega_{12} & -\omega_{02} & \omega_{01} & 0 \end{pmatrix}. \tag{8.19}$$

From (8.17) and (8.19), we obtain

$$\hat{\omega}^{\mu\kappa}\omega_{\kappa\nu} = \Omega g^\mu{}_\nu, \quad \hat{\omega}^{\mu\nu}\omega_{\nu\mu} = 4\Omega. \tag{8.20}$$

According to (8.16), we have

$$(\hat{\omega}^{\mu\kappa}\omega_{\kappa\nu})u^\nu = \Omega g^\mu{}_\nu u^\nu = T\hat{\omega}^{\mu\kappa}\partial_\kappa(\sigma/n), \tag{8.21}$$

thus, we obtain

$$\Omega_T u^\mu = \hat{\omega}^{\mu\kappa}\partial_\kappa(\sigma/n), \quad \text{where} \quad \Omega_T := \Omega/T. \tag{8.22}$$

By substituting (8.19) into (8.22) and with a series of manipulations based on the skew symmetry of $\omega_{\mu\nu}$, we can finally derive

$$\Omega_T \begin{pmatrix} u^0 \\ u^1 \\ u^2 \\ u^3 \end{pmatrix} = \begin{pmatrix} -\partial_1[\omega_{23}(\sigma/n)] - \partial_2[\omega_{31}(\sigma/n)] - \partial_3[\omega_{12}(\sigma/n)] \\ \partial_0[\omega_{23}(\sigma/n)] - \partial_2[\omega_{03}(\sigma/n)] + \partial_3[\omega_{02}(\sigma/n)] \\ \partial_0[\omega_{31}(\sigma/n)] + \partial_1[\omega_{03}(\sigma/n)] - \partial_3[\omega_{01}(\sigma/n)] \\ \partial_0[\omega_{12}(\sigma/n)] - \partial_1[\omega_{02}(\sigma/n)] - \partial_2[\omega_{01}(\sigma/n)] \end{pmatrix}. \quad (8.23)$$

Based on this expression, we can see that

$$\partial_\nu(\Omega_T u^\nu) = 0. \quad (8.24)$$

On the other hand, using (8.16), we have

$$- \Omega u^0 = T[\omega_{23}\partial_1(\sigma/n) + \omega_{31}\partial_2(\sigma/n) + \omega_{12}\partial_3(\sigma/n)], \quad (8.25)$$

thus, we obtain

$$\Omega_T = -[\omega_{23}\partial_1(\sigma/n) + \omega_{31}\partial_2(\sigma/n) + \omega_{12}\partial_3(\sigma/n)]/u^0, \quad (8.26)$$

which is the relativistic expression of Ertel's potential vorticity Q given in (8.5). Following the convention of theoretical physics, we refer to Ω_T as Ertel's charge. The conservation property of (8.24) can be extended to curved space-time by substituting ∂_μ in (8.23) with the covariant derivative ∇_μ and using the tensor identity

$$(\nabla_\mu \nabla_\nu - \nabla_\nu \nabla_\mu)\omega_{\kappa\lambda} = -R^\sigma_{\kappa\mu\nu}\omega_{\sigma\lambda} - R^\sigma_{\lambda\mu\nu}\omega_{\kappa\sigma} \quad (8.27)$$

to calculate the vector divergence on the right side of (8.23), where $R^\alpha_{\beta\gamma\delta}$ denotes the Riemann curvature tensor. According to the second equation in (8.16), the physical dimension of Ω in (8.18), denoted by $dim[\Omega]$, becomes $dim[\Omega] = l^{-2}dim[(w/n)^2]$, where l denotes the length scale. Since n and w are the particle number and the energy per unit volume, respectively, if we use a natural unit system, then $dim[n] = l^{-3}$ and $dim[w^2] = l^{-8}$. Thus, $dim[\Omega] = l^{-1}/l^3$, indicating that $dim[\Omega] = dim[T^{\mu\nu}]$ in (8.11) and hence, from (8.9), *the physical dimension of Ω_T is the entropy per unit volume*.

A particularly intriguing property of $\omega_{\mu\nu}$ is that "Dirac's γ matrix" $\hat{\gamma}^{\mu\nu}$ can be constructed from (8.20). In fact, if we define $\hat{\gamma}^\mu_{\nu}$ such that $\hat{\gamma}^\mu_{\nu} := \hat{\omega}^{\mu\sigma}\omega_{\sigma\nu}$, then, according to (8.20), $\hat{\gamma}^\mu_{\nu} = \Omega g^\mu_{\nu}$. By raising the suffix ν, we have $\hat{\gamma}^{\mu\nu} = \Omega g^{\mu\nu}$; thus, we find that

$$\frac{1}{\Omega}(\hat{\gamma}^{\mu\nu} + \hat{\gamma}^{\nu\mu}) = 2g^{\mu\nu}, \quad (8.28)$$

which is identical to the well-known anti-commutation relation. To further examine the implications of (8.28), we investigated the relation between $g^{\mu\nu}$ and $\hat{\omega}^{\mu\sigma}\omega_\sigma^{\nu}/\Omega$. According to (8.20), $g^{\mu\nu}$ can be rewritten as follows:

$$g^{\mu\nu} = \frac{\hat{\omega}^{\mu\sigma}\omega_\sigma^{\nu}}{\Omega} = \frac{\hat{\omega}^{\mu\sigma}\omega_\sigma^{\nu}(\hat{\omega}^{\kappa\lambda}\omega_{\lambda\kappa})}{\Omega(\hat{\omega}^{\kappa\lambda}\omega_{\lambda\kappa})} = \frac{\hat{\omega}^{\mu\sigma}\omega_\sigma^{\nu}(\hat{\omega}^{\kappa\lambda}\omega_{\lambda\kappa})}{(\hat{\omega}^{\kappa\lambda}\omega_{\lambda\kappa})^2/4}. \quad (8.29)$$

8.2 Converted Form of the Relativistic Equation of Motion and of PV

Recall that, in general, $g^{\mu\nu}$ is not a physical quantity but rather a purely mathematical quantity. However, there exists an exceptional case in which $g^{\mu\nu}$ becomes physical, as shown by (8.30) below, which was derived by lengthy straightforward calculations [6 in Sect. 7] on the Weyl conformal tensor $W_{\alpha\beta\gamma\delta}$.

$$W^{\mu\alpha\beta\gamma} W^{\nu}{}_{\alpha\beta\gamma} - \frac{1}{4} W^2 g^{\mu\nu} = 0, \quad W^2 := W^{\alpha\beta\gamma\delta} W_{\alpha\beta\gamma\delta}. \tag{8.30}$$

(8.30) shows that, for non-vanishing W^2, the cosmological term $\Lambda g^{\mu\nu}$ can be interpreted not as vacuum but as (conformal) gravitational energy. By comparing (8.29) with (8.30), we find that

$$g^{\mu\nu} = \frac{W^{\mu\alpha\beta\gamma} W^{\nu}{}_{\alpha\beta\gamma}}{W^2/4} = \frac{\hat{\omega}^{\mu\sigma} \hat{\omega}^{\kappa\lambda} \omega^{\nu}{}_{\sigma} \omega_{\kappa\lambda}}{(\hat{\omega}^{\kappa\lambda} \omega_{\lambda\kappa})^2/4} = \frac{\hat{\omega}^{\mu\sigma} \hat{\omega}^{\kappa\lambda} \omega^{\nu}{}_{\sigma} \omega_{\kappa\lambda}}{(4\Omega)^2/4}, \tag{8.31}$$

which clearly shows that $(4\Omega)^2$ correlates directly with W^2.

Thus, we have shown that by investigating the relativistic form of Ertel's charge Ω_T, which has been largely ignored except in the field of geophysical fluid dynamics, we found that $\Omega_T u^{\mu}$ is a conserved "entropy current". The importance of this finding is that while the physical dimension of Ω_T is the entropy per unit volume, this quantity is not identical to the thermodynamic entropy density; instead, it is related to both the vortical modes of a given energy-momentum field $T^{\mu\nu}$ and the associated space-time $g^{\mu\nu}$ (defined within the framework of conformal gravity (8.30)).

A possible connection between gravitational entropy and Weyl tensor was suggested by Penrose in terms of *Weyl curvature hypothesis* [6] which may explain the observed extremely isotropic space-time structure of our early universe. In addition to that, we further note that $\Lambda_{dm} g_{\mu\nu}$ with (8.31) is a promising candidate for the ground state of a dark matter model $\Omega_T u^{\mu}$ since it is closely related to the entropy current $\Omega_T u^{\mu}$ which behaves like *a fluid particle field*. In order to support our proposal on the dark matter model, we find that recent studies by Aoki et al. [7] on conserved charges in a curved space-time are quite helpful. Among others, they derived a general form of the definition of a timelike entropy current. The key concept in their definition is, what they call, *timelike intrinsic vector* $\hat{\zeta}^{\mu}$ satisfying

$$T^{\mu}{}_{\nu} \nabla_{\mu} \hat{\zeta}^{\nu} = 0, \quad \Longrightarrow \quad (\nabla_{\mu} \hat{\zeta}^{\mu} = 0, \text{ for } T^{\mu}{}_{\nu} = \Lambda_{dm} g^{\mu}{}_{\nu}), \tag{8.32}$$

where ∇_{μ} and $T^{\mu}{}_{\nu}$ respectively denote covariant derivative and the energy-momentum tensor under consideration, the latter of which becomes $\Lambda_{dm} g^{\mu}{}_{\nu}$ for the ground state of dark matter to be explained in the following subsect. 8.3.

From (8.10) and (8.13), we have the well-known equation: $\nabla_{\nu}(\sigma u^{\nu}) = 0$. Interestingly enough, we have already shown that Ω_T in (8.26) whose physical dimension is exactly the same as that of σ also satisfies the same equation of $\nabla_{\nu}(\Omega_T u^{\nu}) = 0$, though Ω_T is composed of a certain vortical field while σ is not such a kind of quantity. Since the physical meaning of $\nabla_{\nu}(\Omega_T u^{\nu}) = 0$ must be understood in terms of (8.32), we can say that, thanks to it, $\Omega_T u^{\nu}$ is a timelike gravitational entropy current associated with Weyl curvature field. A novel proposal of dark matter model based on this timelike gravitational entropy current will be further explained in the following subchapter.

8.3 On Gravitational Entropy and Dark Matter

In the previous subchapter, we showed that $dim[\Omega] = dim[T^{\mu\nu}]$ in (8.11), and we found that the non-zero value of Ω^2 corresponds to the non-zero value of W^2, which suggests that the non-zero Ω is associated with a special energy field associated with non-zero W^2. In general, in the following Einstein field equation (8.33)

$$R^{\mu\nu} - \frac{1}{2} R g^{\mu\nu} + \Lambda g^{\mu\nu} = -\frac{8\pi G}{c^4} T^{\mu\nu}, \tag{8.33}$$

energy-momentum fields are associated directly with Ricci curvature terms, thus, a peculiar energy field such as Ω would be related to the cosmological term $\Lambda g^{\mu\nu}$ we discussed with (8.30).

As the first step toward understanding the physical meaning of a conserved "entropy" density $\Omega_T (= \Omega/T)$, we introduce a constant reference temperature T_R, the magnitude of which is immaterial in our present discussion, but would become important in the discussion of *low energy states* (8.10) of a relativistic perfect fluid touched upon in subchapter 8.2. With T_R, we can introduce a non-dimensional parameter $\tilde{n} := T_R/T$, which is inversely proportional to the temperature T. With $\tilde{n}$, we can rewrite (8.24) as

$$\nabla_\nu (\tilde{n} \Omega u^\nu) = 0. \tag{8.34}$$

Therefore, if we redefine $\tilde{\Omega}$ as $\tilde{\Omega} := \tilde{n} \Omega$ and introduce $\tilde{T}^{\mu\nu} := \tilde{\Omega} u^\mu u^\nu$, then, we can obtain

$$\nabla_\nu \tilde{T}^{\mu\nu} = 0, \tag{8.35}$$

since u_μ satisfies the geodesic condition $u^\nu \nabla_\nu u_\mu = 0$ at the galactic scale. Note that the non-zero $\tilde{\Omega}$ does not correspond to the non-zero Ricci scalar curvature R, but instead corresponds to the non-zero Weyl curvature W^2; thus, the non-zero current $\tilde{\Omega} u^\mu$ can exist even in "nearly vacuum" regions where $R^{\mu\nu} \approx 0$. Furthermore, since the magnitude of $\tilde{\Omega}$ is inversely proportional to T, the current $\tilde{\Omega} u^\mu$ is a promising candidate for the (cold) dark matter model. Notice that, in the case where $\tilde{\Omega}$ is quantized by $\tilde{\Omega}_0$, namely, $\tilde{\Omega}(n) = n\tilde{\Omega}_0$, then, in (8.33), we have an additional term arising from the ground state of dark matter field, which can be represented either by $\Lambda_{dm} g^{\mu\nu}$ on the l.h.s. or by $(-8\pi G/c^4)\tilde{\Omega}_0 u^\mu u^\nu$ on the r.h.s.

References

1. Ertel, H.: Meteorol. Z. **59**, 277 (1942)
2. Arnol'd, V.I.: Dokl. Akad. Nauk SSSR **162**, 975 (1965). [Sov. Math. Dokl. **6**, 773, (1965)] (in Russian)
3. Holm, D.D., Kuperschmidt, B.A., Levermore, C.D.: Canonical maps between Poisson brackets in Eulerian and Lagrangian descroptions of continuum mechanics. Phys. Lett. A **98**, 389–395 (1983)
4. Kuroda, Y.: Symmetries and Casimir invariants for perfect fluid. Fluid Dynam. Res. **5**, 273–287 (1990)
5. Sakuma, H., Fukumoto, Y.: On formal stability of stratified shear flows. Publ. RIMS Kyoto Univ. **51**, 605 (2015)
6. Penrose, R.: Singularities and time-asymmetry. In: General Relativity: an Einstein Centenary Survey **1**, 581–638 (1979)
7. Aoki, S., Onogi, T., Yokohama, S.: Charge conservation, entropy current and gravitation. Int. J. Mod. Phys. A **36**, 2150098 (2021)

Chapter 9
Novel Cosmology to be Opened up by Off-Shell Science

Abstract In this chapter, we are going to develop a novel cosmological model based on the outcome of *the off-shell science* we have explained so far. We begin our discussion with a brief reviewing of our studies summarized in the following subchapter 9.1 ([i]–[iv]). Since cosmology remains a vast, unknown frontier in the physical sciences, in dealing with mysterious cosmological problems, we adopt a simple and effective approach, the Occam's razor principle. In line with studying the *extended light field*, we employ a couple of unique hypotheses proposed by Petit [12 in Sect. 7] and Penrose [1] called conformal cyclic cosmology (CCC). The first one is on the twin structure (matter and anti-matter) of the universe, while the second one is, as its name shows, on the cyclic temporal evolution of the universe. Since our new cosmological model can be regarded as a variant of CCC, we call it CCC of the second kind (CCC-2nd) and will be explained in subchapter 9.3. According to CCC-2nd, as in the case of the creation and annihilation of the matter and antimatter pair, the twin universes as metric space-time are birthed by a certain kind of SCSB of a light field with null distance ($ds^2 = 0$). The twin universes thus created are divided by an event horizon intrinsically embedded in the de Sitter space structure caused by the most dominant component of dark energy. Eons later, the twins meet at the event horizon to return to the original light field, and this cycle repeats forever. In our cosmological hypothesis, the flatness, isotropy, and horizon problems are resolved, respectively, by the observed existence ratio of dark energy to matter, the Weyl curvature hypothesis proposed by Penrose [6 in Sect. 8], and the existence of a superluminous off-shell electromagnetic field.

9.1 Brief Summary on Our New Studies Explained So Far

[i] The extended electromagnetic 4-vector potential U_μ can be represented by CP with the parameters (λ, ϕ); the former satisfies the spacelike KG equation $\nabla^\sigma \nabla_\sigma \lambda - (\kappa_0)^2 \lambda = 0$, where κ_0 is the experimentally determined dressed photon constant, while the latter satisfies either the same KG equation or $\nabla^\sigma \nabla_\sigma \phi = 0$,

---

© The Author(s), under exclusive license to Springer Nature Switzerland AG 2025
M. Ohtsu and H. Sakuma, *Dressed Photons to Revolutionize Modern Physics*,
Nano-Optics and Nanophotonics, https://doi.org/10.1007/978-3-031-77944-2_9

depending on whether U_μ is spacelike or lightlike. The lightlike U_μ may be interpreted as a $U(1)$ gauge boson, while the spacelike U_μ provides the necessary spacelike momentum supports for field interactions. In fluid mechanics, CP is used for canonical H formulations of barotropic fluids. CP is suitable for extended free Maxwell fields because, in sharp contrast to (8.18) in the baroclinic case, the following pseudo-scalar $\Omega_{(ro)}$ vanishes:

$$\Omega_{(ro)} := S_{01} S_{23} + S_{02} S_{31} + S_{03} S_{12} = 0, \tag{9.1}$$

where $S_{\mu\nu} := \nabla_\mu U_\nu - \nabla_\nu U_\mu$ denotes the field strength of the extended electromagnetic field. According to (9.1), as in the case of a free electromagnetic wave, the extended "electric" and "magnetic" fields are perpendicular to each other.

[ii] The energy-momentum tensor $\hat{T}^{\mu\nu}$ for both cases can be written in a unified form as

$$\hat{T}^{\mu\nu} = \hat{S}^{\mu}{}_{\sigma}{}^{\nu\sigma} - \frac{1}{2} \hat{S}^{\alpha\beta}{}_{\alpha\beta} g^{\mu\nu}, \quad \hat{S}_{\alpha\beta\gamma\delta} := S_{\alpha\beta} S_{\gamma\delta}. \tag{9.2}$$

Note that due to the skew-symmetric nature of $S_{\mu\nu}$, $\hat{S}_{\alpha\beta\gamma\delta}$ satisfies exactly the same properties as the Riemann curvature tensor $R_{\alpha\beta\gamma\delta}$; that is,

$$R_{\beta\alpha\gamma\delta} = -R_{\alpha\beta\gamma\delta}, \quad R_{\alpha\beta\delta\gamma} = -R_{\alpha\beta\gamma\delta}, \quad R_{\gamma\delta\alpha\beta} = R_{\alpha\beta\gamma\delta}, \tag{9.3}$$

$$R_{\alpha\beta\gamma\delta} + R_{\alpha\gamma\delta\beta} + R_{\alpha\delta\beta\gamma} = 0. \tag{9.4}$$

(9.4) is known as the first Bianchi identity and corresponds qualitatively to the second equation in (7.22). Therefore, $\hat{T}^{\mu\nu}$ in (9.2) becomes isomorphic to the Einstein tensor $G^{\mu\nu}$ referred to in (7.44), and its divergence vanishes. *Specifically, we can say that $\hat{T}^{\mu\nu}$ naturally fits into the geometrodynamics of general relativity.*

[iii] Since the non-lightlike U^μ has a spacelike momentum field parameterized by κ_0 in spacelike KG equations (7.36, 7.37), it forms a submanifold of de Sitter space (a pseudo-hypersphere $\mathfrak{D}$ embedded in R^5) explained in subchapter 7.4. According to Sakuma and Ojima [2], the importance of de Sitter space $\mathfrak{D}$ is twofold. Firstly, using a spacelike momentum field on $\mathfrak{D}$, Snyder derived a space-time quantization with a built-in Lorentz invariance. Secondly, de Sitter space is a solution to the Einstein field equation, which describes the accelerated expansion of the universe. In accordance with Snyder's work, they showed that the quantized form of $\hat{T}^{\mu\nu}$ can be given by a combined form of the Majorana fermionic field, which behaves as an energy-momentum tensor with "virtual photons" acting as mediators of electromagnetic field interactions. They further showed that $\hat{T}^{\mu\nu}$ has a unique "ground state" $|M3\rangle_g$ which can be regarded as a compound Rarita-Schwinger state with a spin of $3/2$. In terms of the accelerated expansion of the universe, they considered the possibility that the trace of the energy-momentum tensor representing $|M3\rangle_g$ can be interpreted as a "reduced cosmological constant" Λ_{DP} whose magnitude can be evaluated by the new theory of dressed photons. In fact, Λ_{DP} is found to be $2.47 \times 10^{-53} m^{-2}$,

9.2 On the Meaning of DP Constant

while the observed value of Λ_{obs}, derived from Planck satellite observations [13 in Chap. 7], is $3.7 \times 10^{-53} m^{-2}$.

[iv] In addition to the identification of *physical spacelike part of* space-time, we found that a similar identification of *physical timelike part of* space-time is possible through (8.31) which combines the notion of Ertel's charge in the field of geophysical fluid dynamics with that of Weyl curvature tensor in conformal gravity. Needless to say, metric tensor $g_{\mu\nu}$ itself is not a physical quantity because it is directly related to the choice of coordinate. Thus, we can say that the well-known cosmological term $\Lambda g_{\mu\nu}$ is not physical, which seems to be known as what we call Einstein's "mollusk". The advantage of our new definition of $g_{\mu\nu}$ is twofold. The one is that it is physical in the sense that it is related to Weyl curvature and hence to spin-network. And the other is that it is also related to the entropy of gravitational field through the quantity Ω_T in (8.22).

9.2 On the Meaning of DP Constant

As we explained in item **[iii]**, the observable effect of dark energy is generated by the reduced cosmological constant Λ_{DP}, which has been shown to be proportional to $3(\kappa_0)^2$. The factor 3 in the expression of Λ_{DP} simply reflects the fact that the spatial dimension of our universe is three. Recall that our original goal, which led to the introduction of Λ_{DP}, was to properly evaluate the involvement of the spacelike momentum field in quantum field interactions, and the existence of Λ_{DP} was derived from electromagnetic field interactions as an important parameter characterizing "the ground state" of spacelike virtual photons. Although a similar argument can be extended to gravitational field interactions, we are not going to dwell on it mainly because the magnitude of gravitational field interactions is extremely smaller than that of electromagnetic field interactions and partly because we do not have a satisfactory quantum gravitational theory.

In contrast to the case of dark energy, we can investigate a different possibility for the dark matter model. Recall that the reason why we cannot bring $\hat{T}_\mu{}^\nu$ in (7.41) directly into the right hand side of Einstein field equation (7.44) is because the latter is a timelike field equation while the former is a spacelike one. So, the effect of $\hat{T}_\mu{}^\nu$ on Einstein field equation was put into (7.44) as a parameterized form of $\Lambda_{de} g_\mu{}^\nu$ where $\Lambda_{de} = \Lambda_{DP}$. As was already shown by (8.31), in our newly proposed dark matter model, Weyl curvature plays a key role though Einstein field equation is expressed in terms of Ricci curvature except for the cosmological term having the form of $\Lambda g_\mu{}^\nu$. Therefore, it seems natural to assume that the form $\Lambda_{dm} g_\mu{}^\nu$, where $g_\mu{}^\nu$ is given by (8.31), represents the energy-momentum tensor of the conformal gravity field under the condition that

$$- \Lambda_{dm} \approx \Lambda_{de}/3 > 0, \tag{9.5}$$

which reflects the observationally derived value on the percentage of dark energy and matter.

Using (7.61, 7.63) and (9.5), we have

$$-\Lambda_{dm} = \frac{4\pi Gh(\kappa_0)^2}{c^3\epsilon}, \tag{9.6}$$

which is rewritten as follows in terms of the Planck length l_p, length scales of the universe l_{dm} and DP constant l_{dp}:

$$l_p : = \sqrt{hG/c^3}, \quad l_{dm} := \sqrt{(-\Lambda_{dm})^{-1}}, \quad l_{dp} = (\kappa_0)^{-1}, \tag{9.7}$$

$$(153) \quad \rightarrow \quad l_p l_{dm} = \frac{\sqrt{\epsilon}}{2\sqrt{\pi}} l_{dp} \quad \rightarrow \quad \left[l_p l_{dm} = (\hat{l}_{dp})^2 \right]. \tag{9.8}$$

(9.8) *reveals that if we choose* $\hat{l}_{dp} := l_{dp}/2\sqrt{\pi}$ *as the third component of a natural unit in which we set* $\hat{l}_{dp} = 1$, *then* $\hat{l}_{dp}$ *gives the geometric mean of the smallest scale* l_p *and the largest one of* l_{dm} *in that natural unit system.* By rewriting (7.70) in which we have $\Lambda_{DP} = \Lambda_{de}$ as

$$l_{dp} = \sqrt{\frac{12\pi Gh}{c^3\epsilon}} (\Lambda_{de})^{-1/2}, \quad \rightarrow \quad l_{dp}^\dagger = \sqrt{\frac{12\pi Gh}{c^3\epsilon}} (\Lambda_{obs})^{-1/2}, \tag{9.9}$$

we can use this equation to estimate the DP constant $l_{dp}^\dagger$ solely by the fundamental physical constants G, h, and c together with the observed cosmological constant Λ_{obs} in place of the above Λ_{de}. Directly from the second equation in (9.9), we obtain

$$l_{dp}^\dagger \approx 40.0 \text{ nm}, \quad [\text{Experiments}: 50 \text{ nm} < l_{dp} < 70 \text{ nm}]. \tag{9.10}$$

9.3 Twin Structure of the Universe

In subchapter 7.4, we have already touched on de Sitter space given by (7.68), which is the solution to Einstein field equation of the form:

$$R_\mu{}^\nu - \frac{R}{2} g_\mu{}^\nu + \Lambda_{de} g_\mu{}^\nu = 0. \tag{9.11}$$

Using (9.5), (7.68) can be rewritten as

$$ds^2 = (cdt)^2 - (R_0)^2 \exp[2\sqrt{-\Lambda_{dm}} ct][dr^2 + r^2(d\theta^2 + \sin^2\theta)d\varphi^2]. \tag{9.12}$$

Since the concrete form of a solution to a given differential equation depends on the choice of coordinates, the solution to (9.11) would have a form different from

9.3 Twin Structure of the Universe

(9.12) through coordinate transformation. Among others, especially intriguing one is the following coordinate transformation

$$l_p r = \frac{r'}{\sqrt{D}} \exp[-\sqrt{-\Lambda_{dm}} ct'], \quad t = t' + \frac{1}{2c}\sqrt{\frac{1}{-\Lambda_{dm}}} \ln D, \quad (9.13)$$

where D is defined either (case I) by $1 > D := 1 + \Lambda_{dm}(r')^2 > 0$, or (case II) by $1 > D := -\Lambda_{dm}(r')^2 - 1 > 0$. With this coordinate transformation, (9.12) turns into the following "stationary" metric

$$ds^2 = \left(1 + \Lambda_{dm}(r')^2\right)(cdt')^2 - \frac{(dr')^2}{\left(1 + \Lambda_{dm}(r')^2\right)} - (r')^2(d\theta^2 + \sin^2\theta d\varphi^2). \quad (9.14)$$

Note that the metric (9.14) is similar in form to Schwarzschild metric given below, for which an event horizon exists at $r' = \alpha$, while that in (9.14) exists at $r' = \sqrt{1/-\Lambda_{dm}}$.

$$ds^2 = \left(1 - \frac{\alpha}{r'}\right)(cdt')^2 - \frac{(dr')^2}{\left(1 - \frac{\alpha}{r'}\right)} - (r')^2(d\theta^2 + \sin^2\theta d\varphi^2). \quad (9.15)$$

In case I of the stationary metric (9.14), using (9.13) we first see that $r' = 0$ corresponds to $r = 0$ and $t' = t$. And we further see that there exists a positive parameter $\Theta > 1$ which connects t and t' through $t' = \Theta t$, with which the second equation in (9.13) is rewritten as

$$(\Theta - 1)t = -\frac{1}{2c}\sqrt{\frac{1}{-\Lambda_{dm}}} \ln D > 0. \quad (9.16)$$

Using this, we see that r' moves from 0 to $1/\sqrt{(-\Lambda_{dm})}$ as t moves from 0 to $+\infty$. Similarly, in case II, we see that r' moves from $\sqrt{2/(-\Lambda_{dm})}$ to $1/\sqrt{(-\Lambda_{dm})}$ as t moves from 0 to $+\infty$. This dual structure, illustrated in Fig. 9.1, clearly shows that by taking $t = 0$ as the origin of time from which twin Big Bang universes evolve, they will meet at the event horizon in (9.14) after an eon later ($t = \infty$).

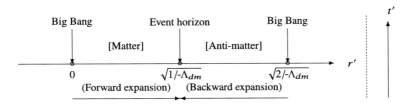

Fig. 9.1 Dual configuration of twin universes

To the best of our knowledge, the concept of twin universes with matter vs. anti-matter duality was first discussed by Petit [12 in Chap. 7]. We think that his basic idea on cosmological model fits into the configuration illustrated in Fig. 9.1, which tells us that $\sqrt{(-\Lambda_{dm})^{-1}}$ is the genuine characteristic length scale of our universe. This justifies the fact that Λ_{dm} defined in (9.5) is the cosmological constant that appears in Einstein field equation (8.33). The forward and backward time evolutions of twin universes correspond, respectively, to positive and negative field operators of the 4-momentum, while the existence of twin universes naturally explains the reason why one-sided energy spectra at the level of state vector space works for many practical situations in each universe. If the birth of these twin universes was brought about by conformal symmetry breaking of certain light fields in which the duality between "matter (with positive energy) and antimatter (with negative energy)" works as the separation rule of the twin structure, then the twin pair will return to the original light fields when they meet at the event horizon. The next Big Bangs of the twin pair will occur at certain locations on this event horizon distant from each other by $\sqrt{2/-\Lambda_{dm}}$.

9.4 On Spontaneous Conformal Symmetry Breaking of Light Fields

In our cosmological scenario of CCC-2nd, $\hat{T}_\mu{}^\nu$ in (7.41) shown to be isomorphic to the spacelike Einstein tensor $G_\mu{}^\nu$ and $\Lambda_{dm}g^{\mu\nu}$ with $g^{\mu\nu}$ given by (8.31), provide *the physical* spacelike and timelike space-times, respectively. These space-times emerged from the spontaneous conformal symmetry breaking (SCSB) of the *primordial* light field with null distance ($ds^2 = 0$). To consolidate our CCC-2nd scenario here, we show that a unique vector boson exists in our model, which plays the role of Nambu–Goldstone boson (NGB) associated with the SCSB process.

To start with, we consider again the spacelike CP of U^μ explained in (7.36, 7.37 and 7.38). The field strength $S_{\mu\nu}$ defined by the curl of U_μ assumes the same form as the one in the lightlike case given in (7.33). The null geodesic equation: $U^\nu \nabla_\nu U_\mu = 0$ assumes the form of

$$U^\nu \nabla_\nu U_\mu = -S_{\mu\nu}U^\nu + \nabla_\nu (U^\nu U_\nu/2) = 0; \quad U^\nu U_\nu < 0. \tag{9.17}$$

In relativistic fluid dynamics, the magnitude of U^μ, defined by $V := U^\nu U_\nu/2$, can be normalized as $V = 1$ [1 in Chap. 7]. As the equations on λ and ϕ in (7.36, 7.37) are linear, we can introduce a similar normalization for $L_\mu = \nabla_\mu \lambda$ and $C_\mu = \nabla_\mu \phi$. The natural normalization would be

$$L^\nu L_\nu = -(m_\lambda)^2 \lambda^2, \quad C^\nu C_\nu = -(m_\phi)^2 \phi^2, \tag{9.18}$$

where the wavenumber vector, k^μ, of the respective plane-wave solutions satisfies $k^\nu k_\nu = -(m_\lambda)^2$ and $k^\nu k_\nu = -(m_\phi)^2$. For the spacelike CP of U^μ discussed thus far, we have

$$(m_\lambda)^2 = (m_\phi)^2 = (\kappa_0)^2. \qquad (9.19)$$

Using the orthogonality condition, $C^\nu L_\nu = 0$ in (7.38), and (9.18), a couple of important characteristics of spacelike U^μ fields can readily be expressed as

$$\nabla_\nu U^\nu = 0, \quad V = -\frac{1}{8}\lambda^2\phi^2[(m_\lambda)^2 + (m_\phi)^2]. \qquad (9.20)$$

Now, revisiting (7.36, 7.37), we consider a different case in which we only replace the second equation, $\nabla^\nu\nabla_\nu\phi - (\kappa_0)^2\phi = 0$, with $\nabla^\nu\nabla_\nu\phi + (\kappa_0)^2\phi = 0$. With this change, (9.19) changes into $(m_\lambda)^2 = (\kappa_0)^2 = -(m_\phi)^2$. Consequently, V in (9.20) vanishes, and U^μ becomes a null vector. Notably, the form of $S_{\mu\nu}U^\nu$ in (9.17) is similar to that of Lorentz force, $F_{\mu\nu}(ev^\nu)$, where $F_{\mu\nu}$ and ev^ν denote the background electromagnetic field and an electric current with charge e, respectively. A direct expression of $-S_{\mu\nu}U^\nu$ becomes

$$\begin{aligned} -S_{\mu\nu}U^\nu &= -(L_\mu C_\nu - L_\nu C_\mu)(\lambda C^\nu - \phi L^\nu)/2 \\ &= -\lambda\phi(\kappa_0)^2(\phi L_\mu - \lambda C_\mu)/2 = \lambda\phi(\kappa_0)^2 U_\mu. \end{aligned} \qquad (9.21)$$

Alternatively, as we have $\nabla_\nu U^\nu = -\lambda\phi(\kappa_0)^2$, in this case, (9.21) becomes

$$S_{\mu\nu}U^\nu = -\lambda\phi(\kappa_0)^2 U_\mu = (\nabla_\nu U^\nu)U_\mu \neq 0. \qquad (9.22)$$

(9.22) expresses that *the null vector U^μ with a non-vanishing irrotational part* is a vector field for which κ_0 and $i\kappa_0$ play the role of the field source as in the case of $\pm e$ in electromagnetism. Furthermore, U^μ and the field source behaves as if it is a "virtual photon excited by electrical charges $\pm e$ responsible for nonlinear electromagnetic field interactions. In our CCC-2nd scenario, the emergence of a couple of these sources (κ_0 and $i\kappa_0$) is interpreted as the consequence of the SCSB of the light vector field whose temporal and spatial components exist in a balanced manner. Thus, we believe that the vector field, U^μ, with the abovementioned characteristics plays the role of NGB in our cosmological scenario. Moreover, it plays substantial roles in the nonlinear field interactions between the quantum generalized sectors defined by Ojima [4 in Chap. 5], existing on not only timelike but also spacelike domains, and the classical generalized sectors on timelike domains.

References

1. Penrose, R.: Before the Big Bang: an outrageous new perspective and its implications for particle physics. Proc, EPAC (2006)
2. Sakuma, H., Ojima, I.: On the dressed photon constant and its implication for a novel perspective on cosmology. Symmetry **13**, 593 (2021). https://doi.org/10.3390/sym13040593

Chapter 10
Implications of the Novel Cosmology

Abstract In this final chapter, in order to demonstrate the validity of newly proposed cosmological theory explained in the previous Chap. 9, we will show that an intriguing clue to the longstanding hierarchy problem is attained through our new models of dark energy and matter together with the important cosmological meaning of DP constant from the viewpoint of micro-quantum and macro-classical sector theory introduced by Ojima's MMD theory referred to in subchapter 5.1. In doing so, we will also touch on a couple of least expected similarities between the outcomes of our cosmological model and those of superstring theories discussed respectively by Witten and Maldacena.

10.1 On Hierarchy Problem in Particle Physics

Based on what we have explained in Chaps. 7 and 8. we now see that Einstein field equation incorporating our dark energy and matter models is expressed as follows:

$$R_\mu{}^\nu - \frac{R}{2}g_\mu{}^\nu + \Lambda_{dm}g_\mu{}^\nu + \Lambda_{de}g_\mu{}^\nu = -\frac{8\pi G}{c^4}T_\mu{}^\nu,$$
$$-\Lambda_{dm} = \Lambda_{de}/3 + \epsilon > 0. \tag{10.1}$$

Our present concern in this section is the dual gravitational field of the dark matter field mentioned on p.101 : $\Lambda_{dm}g_\mu{}^\nu$, which should not be confused the isotropic cosmological term. As the first step, by comparing Coulomb's law with the universal law of gravitation, we obtain

$$F_e = \frac{1}{4\pi\epsilon_0}\frac{q_1q_2}{r^2}, \quad F_g = G\frac{m_1m_2}{r^2}, \tag{10.2}$$

where notations are quite conventional. Now, let us examine the ratio F_e/F_g. In doing so, it would be natural to choose the fundamental electric charge (e) for $q_1 = q_2$. However, for $m_1 = m_2$, we encounter a serious challenge as we cannot single out "the fundamental mass charge" like e in the case of F_e. To overcome this problem,

© The Author(s), under exclusive license to Springer Nature Switzerland AG 2025　113
M. Ohtsu and H. Sakuma, *Dressed Photons to Revolutionize Modern Physics*,
Nano-Optics and Nanophotonics, https://doi.org/10.1007/978-3-031-77944-2_10

114   10 Implications of the Novel Cosmology

recall first that, for our dark energy model, a unique state, $|M3\rangle_g$ explained along with (7.56), exists, which behaves like "the ground state" of $S_{\mu\nu}$. And we assume that this "ground state" of dark energy is related to Λ_{dm} of dark matter through (9.5).

Furthermore, using (8.31), we pointed out the existence of the minimum value of W^2, i.e., $(W_0)^2 \neq 0$, associated with the SCSB of the light field. Thus, we may naturally assume that $-\Lambda_{dm} = |W_0|$. Actually, in the early stage of our studies where we did not consider the possibility of the "temporal change" of Λ_{dm}, we simply asserted $-\Lambda_{dm} = W_0 = const$. However, in the present discussion where we consider the "temporal change" of Λ_{dm}, we assume that W_0 is a certain positive constant that satisfies

$$W_0 := Min\{-\Lambda_{dm}\}. \tag{10.3}$$

Under this new hypothesis, *we can regard* $-\Lambda_{dm}$ *as the "temporally changing" fundamental mass of the gravitational field* of which justification will be provided shortly.

As the dimension of Λ_{dm} is $(length)^{-2}$, we introduce m_λ having the dimension of mass, corresponding to Λ_{dm}. Substituting e and m_λ into F_e and F_g in (10.2), respectively, we obtain

$$\frac{F_e}{F_g} = \frac{e^2}{4\pi\epsilon_0}\frac{1}{G(m_\lambda)^2} = \frac{e^2}{4\pi\epsilon_0 c\hbar}\frac{c\hbar}{G(m_\lambda)^2} = \alpha\left(\frac{m_p}{m_\lambda}\right)^2, \tag{10.4}$$

$$\alpha := \frac{e^2}{4\pi\epsilon_0 c\hbar}, \quad m_p := \sqrt{\frac{c\hbar}{G}}, \tag{10.5}$$

where α and m_p are the coupling constant of the electromagnetic field and Planck's mass. For Λ_{dm}, using (9.5), namely, $-\Lambda_{dm} \approx \Lambda_{de}/3$, and using the concrete expression of (7.70), we have [1]

$$-\Lambda_{dm} \approx \frac{4\pi G h}{c^2}\frac{(\kappa_0)^2}{\epsilon} = \frac{8\pi^2 G\hbar}{c^3}\frac{(\kappa_0)^2}{\epsilon} = 8\pi^2(l_p)^2\frac{1}{\epsilon(l_{dp})^2}; \quad l_{dp} := (\kappa_0)^{-1}. \tag{10.6}$$

Here, l_p and l_{dp} are the Planck length and DP length defined as the inverse of κ_0 in (7.30); ϵ denotes a dimension adjusting coefficient of *unit length squared*. Two reasons exist for the appearance of ϵ in the expression of the above Λ_{dm}. First, the quantity Λ_{de}, in (7.70) is related to the "radiation pressure" of the $S_{\mu\nu}$ field. Second, the calculation of such a quantity is required to make the energy quantization of the lightlike $S_{\mu\nu}$ field consistent with $E = h\nu$ for the usual radiation field. As we have introduced m_λ as the elemental mass corresponding to $-\Lambda_{dm}$, we can determine it by the following formal identification using Einstein's field equation:

$$-\Lambda_{(dm)}g_\mu{}^\nu = \frac{8\pi G}{c^4}[(\rho_\lambda c^2)u_\mu u^\nu], \quad \Rightarrow \quad -\Lambda_{(dm)}(l_\epsilon)^3 = \frac{2\pi G}{c^2}m_\lambda, \quad m_\lambda = \rho_\lambda(l_\epsilon)^3, \tag{10.7}$$

10.1 On Hierarchy Problem in Particle Physics 115

where l_ϵ denotes unit length. Therefore, using (10.4), (10.5), (10.6), and (10.7), we finally obtain

$$\frac{F_e}{F_g} = \frac{\alpha}{\pi^2} \frac{(l_{dp})^4}{(l_p)^2} \frac{1}{\epsilon}. \tag{10.8}$$

Substituting $\alpha = 7.3 \times 10^{-3}$, $\pi^2 = 9.9$, $l_p = 1.6 \times 10^{-35}$m, $l_{dp} \approx 5.0 \times 10^{-8}$m, and $\epsilon = 1$m^2 into (10.8), we obtain

$$\frac{F_e}{F_g} = 1.1 \times 10^{37}, \tag{10.9}$$

which appears to be consistent with the conventional rough estimates obtained without using Λ_{dm}.

Having derived (10.8), now we examine the consequence of the "temporal change" of $R_{dS} (= 1/\sqrt{-\Lambda_{dm}})$ applicable to F_e/F_g in (10.8). In theoretical physics, an intriguing possibility that physical constants such as light speed c, gravitational constant G etc. are not constant but are time-dependent variables has been investigated in order to explain the huge difference between F_e and F_g shown in (10.9). Suppose that Λ_{dm} is such a time-dependent quantity, then $\Lambda_{dm} g_\mu{}^\nu$ in (10.1) ceases to be a divergence-free term. Notice, however, that as Λ_{dm} is directly related to the radius R_{dS} (through (9.5)) of de Sitter space (7.68), the partial derivative of Λ_{dm} with respect to the cosmological coordinate (x^μ) vanishes under the assumption that the shape of the isotropic universe is given by that of de Sitter space and (x^μ) is defined on it. This is because $\nabla_\mu R_{dS}$ $(0 \leq \mu \leq 3)$ is on the tangent hyperplane of $R_{dS} = const.$, on which the gradient of the local 4D spacetime coordinate (x^μ) exists. Thus, in this sense, $\Lambda_{dm} g_\mu{}^\nu$ remains a divergence-free term, although the radius can either expand or shrink in the fifth dimension perpendicular to the gradient of the 4D coordinates (x^μ). The "temporal change" of Λ_{dm} that we consider is the change in the magnitudes of the dark energy and matter fields.

For simplicity, in our discussion, except for Λ_{de} and the related Λ_{dm}, we assume that all physical constants are fixed quantities. From (9.5, 10.6), we readily obtain

$$l_{dp} \approx \frac{2\sqrt{2}\pi}{\sqrt{\epsilon}} \frac{l_p}{\sqrt{-\Lambda_{dm}}} = \frac{2\sqrt{2}\pi}{\sqrt{\epsilon}} l_p R_{dS}, \tag{10.10}$$

which implies that the DP length (l_{dp}) affords the geometric mean of the smallest Planck length (l_p) and the largest scale of our universe (R_{dS}). Furthermore, under the assumption that $l_p = const.$, l_{dp} becomes proportional to R_{dS}. Thus, we can choose l_{dp} as the unique geometrical parameter of our cosmological model. From the viewpoint of the unification of four forces, examining the case where we have $F_e/F_g = 1$ is interesting. A simple calculation shows that

$$\text{the present value}: \quad \frac{F_e}{F_g} = 1.1 \times 10^{37}; \quad l_{dp} \approx 5.0 \times 10^{-8}\text{m} \tag{10.11}$$

the unification value : $\dfrac{F_e}{F_g} = 1$; $\qquad l_{dp} \approx 2.4 \times 10^{-17} \text{m}.$ (10.12)

Thus, using (10.6), we observe that the unification value of $-\Lambda_{dm}(u)$ is (4.4×10^{18}) times larger than the present value of $-\Lambda_{dm}(p)$.

We focus on dark energy and matter mainly because of their extreme predominance over material substances. This implies that, in thermodynamic terminology, the dark matter field $(\Lambda_{dm} g_{\mu\nu})$ with negative Λ_{dm} resulting from *Weyl curvature* and dark energy field $-\Lambda_{de} g_{\mu\nu}$ with positive Λ_{de} resulting from *Ricci curvature* would work as high- and low-temperature reservoirs, respectively, for the gravity-driven temporal evolution of such material systems as stars, galaxies, clusters of galaxies, and the large-scale structure of the cosmos. Moreover, in this cosmic thermodynamical system, $\Omega_T u^{\mu}$, defined in (8.26), gives the gravitational entropy flow. As the energy density of the two thermal reservoirs are finite, the initial "temperature difference" between them, measured by $|\Lambda_{dm}(u) - \Lambda_{de}(u)| \approx 4|\Lambda_{dm}| = 4[R_{dS}(u)]^{-2}$, would decrease with the temporal evolution of material systems. This result can be observed as the extra expansion (increase of R_{dS}) of our universe in the fifth dimension, directly related to the temporal increase in the ratio of the coupling constants, $F_e/F_g \propto (R_{dS})^2$ [1].

Although the dynamics we have discussed are unrelated to superstring theories, it is interesting to bring our attention to Witten's noteworthy remark [2] made at *Strings '95*: "eleven-dimensional supergravity arises as a low energy limit of the ten-dimensional Type IIA superstring." This appears to be qualitatively similar to our present situation, in which our $4D$ universe undergoes an extra expansion into the surrounding fifth-dimensional space. The expansion starts from the initial high-energy state of $F_e/F_g \approx 1$ with a negligible magnitude of W^2 in (8.30) to lower energy states having large values of W^2 as the measure of conformal gravity.

10.2 On Maldacena (AdS/CFT) Duality

In subchapter 7.4, we saw that $S_{\mu\nu}$ is closely related to de Sitter space $\mathfrak{D}$ having the well-known scale-free property. Recall first that spinor is an irreducible representation of the universal covering group, $SU(2)$ of $SO(3)$. For the $4D$ spacetime case, in which we have the Lorentz transformation group $(SO(1, 3))$, the Lorentzian spinor $(SL(2, C))$ corresponds to $SU(2)$ in the case of $SO(3)$. When we further extend $SO(1, 3)$ into the $4D$ conformal transformation group $(SO(2, 4))$, $SL(2, C)$ is extended into $SU(2, 2)$, which operates on Penrose's twistor in the $4D$ complex space-time. As in the case of the above extension of spinor, we can also consider a similar extension of the electromagnetic field $(F_{\mu\nu})$ as the $U(1)$ gauge field. We believe that the CP (explained in subchapter 7.2) applied to extend $F_{\mu\nu}$ into a spacelike momentum domain is what is required for such an extension of conformal transformation, which is closely related to the important notion of modular form.

10.2 On Maldacena (AdS/CFT) Duality

The fact that the emergence of de Sitter space through CP is an inevitable consequence of the extension of $SL(2, C)$ can be readily verified from the following properties of the Lorentzian spinor, $\Psi(V^\mu)$:

$$\Psi(V^\mu) = V^{\mu(\mu)'} = \begin{bmatrix} V^{00'} & V^{01'} \\ V^{10'} & V^{11'} \end{bmatrix} = \frac{1}{\sqrt{2}} \begin{bmatrix} V^0 + V^3 & V^1 + iV^2 \\ V^1 - iV^2 & V^0 - V^3 \end{bmatrix}, \quad (10.13)$$

$$det\,\Psi(V^\mu) = \frac{1}{2}[(V^0)^2 - (V^1)^2 - (V^2)^2 - (V^3)^2]. \quad (10.14)$$

For the spacelike vector V^μ, $det\,\Psi(V^\mu)$ becomes negative so that (10.14) becomes isomorphic to the second equation in (7.73). As (7.72) and the second (7.73) are connected by a one-to-one correspondence through parameter η_4 and the latter is isomorphic to (10.14), we thus see that de Sitter space $\mathfrak{D}$ and $\Psi(V^\mu)$ share the same symmetry.

Now, let us compare the following forms of $g^{\mu\nu}$:

$$g^{\mu\nu}_{(I)} = \frac{-1}{\Lambda_{(I)}} \left(R^{\mu\nu}_{(I)} - \frac{R_{(I)}}{2} g^{\mu\nu}_{(I)} \right), \quad (W^{\alpha\beta\gamma\delta} = 0), \quad (I = 1 \text{ or } 2), \quad (10.15)$$

$$g^{\mu\nu} = \frac{W^{\mu\alpha\beta\gamma} W^\nu_{\ \alpha\beta\gamma}}{W^2/4} \quad (W^2 \neq 0), \quad (10.16)$$

where (10.15) represents either de Sitter space with $\Lambda_1 = \Lambda_{de} > 0$ or anti de Sitter space (AdS) with $\Lambda_2 = \Lambda_{dm} < 0$, depending on the index (I).

In our CCC-2nd hypothesis, the isotropy of an early universe is explained by the small amplitude of W^2 (Weyl curvature hypothesis proposed by Penrose [6 in Chap. 8]), which motivates us to examine the possibility that the following limit exists:

$$g^{\mu\nu} = \frac{W^{\mu\alpha\beta\gamma} W^\nu_{\ \alpha\beta\gamma}}{W^2/4} \rightarrow g^{\mu\nu}_{(2)} = \frac{-1}{\Lambda_{(2)}} \left(R^{\mu\nu}_{(2)} - \frac{R_{(2)}}{2} g^{\mu\nu}_{(2)} \right), \text{ as } W^2 \rightarrow 0.$$

$$(10.17)$$

Within the framework of our CCC-2nd hypothesis, this gives a diffeomorphism defined as the time reversal of a given cosmological time development. As $g^{\mu\nu}$ on the l.h.s. gives the gravitational field, whereas the r.h.s. represents AdS, (10.17) can be regarded as the "Maldacena (AdS/CFT) duality" [3] in CCC-2nd.

In our cosmological theory, as (9.5) shows, the universe expands, keeping a quasi-equilibrium between Λ_{dm} and Λ_{de}, which behave as "high and low temperature reservoirs", respectively, for the temporal evolutions of systems in the universe. Thus, the essential global aspect of the expanding universe in a quasi-equilibrium state should be described by the Tomita-Takesaki theory [12 in Chap. 7] as a thermodynamic Kubo-Martin-Schwinger (KMS) state with infinite degrees of freedom. As the KMS state is a mixed one, its corresponding Gel'fand-Naimark-Segal representation is reducible. Therefore, for $\mathcal{M}$, defined as a von Neumann algebra on Hilbert space $\mathfrak{H}$, there exists its commutant, $\mathcal{M}'$, which satisfies the following inversion relation:

$$JMJ = M', \quad e^{itH}Me^{-itH} = M, \quad J^2 = 1, \tag{10.18}$$

$$JHJ = -H, \tag{10.19}$$

where H and J denote the Hamiltonian and anti-unitary operators called modular conjugation, respectively. Notice that the spectrum of the Hamiltonian is symmetric with respect to its sign, which indicates the existence of states with negative energy. We believe that the result of the Tomita-Takesaki theory applies to the case of a twin universe configuration and to the thermodynamics of the observed cosmic background radiation whose energy-spectrum distribution is given by the black body radiation.

References

1. Sakuma, H., Ojima, I., Okamura, K.: Reexamination of the hierarchy problem from a novel geometric perspective, arXiv: 2406.01626
2. Witten, E.: String theory dynamics in various dimension (1995), arXiv:hep-th/9503124v2
3. Maldacena, J.: The large N limit of superconformal field theories and supergravity. Adv. Theor. Math. Phys. **2**, 231–252 (1998)